Computer Algebra and Symbolic Computation

Computer Algebra and Symbolic Computation
Elementary Algorithms

Joel S. Cohen

Department of Computer Science
University of Denver

A K Peters
Natick, Massachusetts

Editorial, Sales, and Customer Service Office
A K Peters, Ltd.
63 South Avenue
Natick, MA 01760
www.akpeters.com

Library of Congress Cataloging-in-Publication Data
Cohen, Joel S.
    Computer algebra and symbolic computation : elementary
algorithms / Joel S. Cohen
        p. cm.
    Includes bibliographical references and index.
    ISBN 1-56881-158-6
    1. Algebra–Data processing. I. Title.

QA155.7.E4 .C635 2002
512–dc21   2001055159

Printed in Canada
06 05 04 03 02   10 9 8 7 6 5 4 3 2 1

To my wife Kathryn

Contents

Preface

Computer algebra is the field of mathematics and computer science that is concerned with the development, implementation, and application of algorithms that manipulate and analyze mathematical expressions. This book and the companion text, *Computer Algebra and Symbolic Computation: Mathematical Methods*, are an introduction to the subject that addresses both its practical and theoretical aspects. This book, which addresses the practical side, is concerned with the formulation of algorithms that solve symbolic mathematical problems, and with the implementation of these algorithms in terms of the operations and control structures available in computer algebra programming languages. *Mathematical Methods*, which addresses more theoretical issues, is concerned with the basic mathematical and algorithmic concepts that are the foundation of the subject. Both books serve as a bridge between texts and manuals that show how to use computer algebra software and graduate level texts that describe algorithms at the forefront of the field.

These books have been in various stages of development for over 15 years. They are based on the class notes for a two-quarter course sequence in computer algebra that has been offered at the University of Denver every other year for the past 16 years. The first course, which is the basis for *Elementary Algorithms*, attracts primarily undergraduate students and a few graduate students from mathematics, computer science, and engineering. The second course, which is the basis for *Mathematical Methods*, attracts primarily graduate students in both mathematics and computer science. The course is cross-listed under both mathematics and computer science.

Prerequisites

The target audience for these books includes students and professionals from mathematics, computer science, and other technical fields who would like to know about computer algebra and its applications.

In the spirit of an introductory text, we have tried to minimize the prerequisites. The mathematical prerequisites include the usual two year freshman–sophomore sequence of courses (calculus through multivariable calculus, elementary linear algebra, and applied ordinary differential equations). In addition, an introductory course in discrete mathematics is recommended because mathematical induction is used as a proof technique throughout. Topics from elementary number theory and abstract algebra are introduced as needed.

On the computer science side, we assume that the reader has had some experience with a computer programming language such as Fortran, Pascal, C, C++, or Java. Although these languages are not used in these books, the skills in problem solving and algorithm development obtained in a beginning programming course are essential. One programming technique that is especially important in computer algebra is recursion. Although many students will have seen recursion in a conventional programming course, the topic is described in Chapter 5 of *Elementary Algorithms* from a computer algebra perspective.

Realistically speaking, while these prerequisites suffice in a formal sense for both books, in a practical sense there are some sections as the texts progress where greater mathematical and computational sophistication is required. Although the mathematical development in these sections can be challenging for students with the minimum prerequisites, the algorithms are accessible, and these sections provide a transition to more advanced treatments of the subject.

Organization and Content

Broadly speaking, these books are intended to serve two (complementary) purposes:

- *To provide a systematic approach to the algorithmic formulation and implementation of mathematical operations in a computer algebra programming language.*

Algorithmic methods in traditional mathematics are usually not presented with the precision found in numerical mathematics or conventional computer programming. For example, the algorithm for the expansion of products and powers of polynomials is usually given informally instead of with (recursive) procedures that can be expressed as a computer program.

The material in *Elementary Algorithms* is concerned with the algorithmic formulation of solutions to elementary symbolic mathematical problems. The viewpoint is that mathematical expressions, represented as expression trees, are the data objects of computer algebra programs, and by using a few primitive operations that analyze and construct expressions, we can implement many elementary operations from algebra, trigonometry, calculus, and differential equations. For example, algorithms are given for the analysis and manipulation of polynomials and rational expressions, the manipulation of exponential and trigonometric functions, differentiation, elementary integration, and the solution of first order differential equations. Most of the material in this book is not found in either mathematics textbooks or in other, more advanced computer algebra textbooks.

- *To describe some of the mathematical concepts and algorithmic techniques utilized by modern computer algebra software.*

For the past 35 years, the research in computer algebra has been concerned with the development of effective and efficient algorithms for many mathematical operations including polynomial greatest common divisor (gcd) computation, polynomial factorization, polynomial decomposition, the solution of systems of linear equations and multivariate polynomial equations, indefinite integration, and the solution of differential equations. Although algorithms for some of these problems have been known since the nineteenth century, for efficiency reasons they are not suitable as general purpose algorithms for computer algebra software. The classical algorithms are important, however, because they are much simpler and provide a context to motivate the basic algebraic ideas and the need for more efficient approaches.

The material in *Mathematical Methods* is an introduction to the mathematical techniques and algorithmic methods of computer algebra. Although the material in this book is more difficult and requires greater mathematical sophistication, the approach and selection of topics is designed so that it is accessible and interesting to the intended audience. Algorithms are given for basic integer and rational number operations, automatic (or default) simplification of algebraic expressions, greatest common divisor calculation for single and multivariate polynomials, resultant computation, polynomial decomposition, polynomial simplification with Gröbner bases, and polynomial factorization.

Topic Selection

The author of an introductory text about a rapidly changing field is faced with a difficult decision about which topics and algorithms to include in the work. This decision is constrained by the background of the audience, the mathematical difficulty of the material and, of course, by space limitations. In addition, we believe that an introductory text should really be an introduction to the subject that describes some of the important issues in the field but should not try to be comprehensive or include all refinements of a particular topic or algorithm. This viewpoint has guided the selection of topics, choice of algorithms, and level of mathematical rigor.

For example, polynomial gcd computation is an important topic in *Mathematical Methods* that plays an essential role in modern computer algebra software. We describe classical Euclidean algorithms for both single and multivariate polynomials with rational number coefficients and a Euclidean algorithm for single variable polynomials with simple algebraic number coefficients. It is well known, however, that for efficiency reasons, these algorithms are not suitable as general purpose algorithms in a computer algebra system. For this reason, we describe the more advanced subresultant gcd algorithm for multivariate polynomials but omit the mathematical justification, which is quite involved and far outside the scope and spirit of these books.

One topic that is not discussed is the asymptotic complexity of the time and space requirements of algorithms. Complexity analysis for computer algebra, which is often quite involved, uses techniques from algorithm analysis, probability theory, discrete mathematics, the theory of computation, and other areas that are well beyond the background of the intended audience. Of course, it is impossible to ignore efficiency considerations entirely and, when appropriate, we indicate (usually by example) some of the issues that arise. A course based on *Mathematical Methods* is an ideal prerequisite for a graduate level course that includes the complexity analysis of algorithms along with recent developments in the field[1].

Chapter Summaries

A more detailed description of the material covered in these books is given in the following chapter summaries.

Elementary Algorithms

Chapter 1: Introduction to Computer Algebra. This chapter is an introduction to the field of computer algebra. It illustrates both the possibilities and limitations for computer symbolic computation through dialogues with a number of commercial computer algebra systems.

---

[1] A graduate level course could be based on one of the following books: Akritas [2], Geddes, Czapor, and Labahn [39], Mignotte [66], Mignotte and Ştefănescu [67], Mishra [68], von zur Gathen and Gerhard [96], Winkler [101], Yap [105], or Zippel [108].

Chapter 2: Elementary Concepts of Computer Algebra. This chapter introduces an algorithmic language called *mathematical pseudo-language* (or simply MPL) that is used throughout the books to describe the concepts, examples, and algorithms of computer algebra. MPL is a simple language that can be easily translated into the structures and operations available in modern computer algebra languages. This chapter also includes a general description of the evaluation process in computer algebra software (including automatic simplification), and a case study which includes an MPL program that obtains the change of form of quadratic expressions under rotation of coordinates.

Chapter 3: Recursive Structure of Mathematical Expressions. This chapter is concerned with the internal tree structure of mathematical expressions. Both the conventional structure (before evaluation) and the simplified structure (after evaluation and automatic simplification) are described. The structure of automatically simplified expressions is important because all algorithms assume that the input data is in this form. Four primitive MPL operators (*Kind, Operand, Number_of_operands,* and *Construct*) that analyze and construct mathematical expressions are introduced. The chapter also includes a description of four MPL operators (*Free_of, Substitute, Sequential_substitute,* and *Concurrent_substitute*) which depend only on the tree structure of an expression.

Chapter 4: Elementary Mathematical Algorithms. In this chapter we describe the basic programming structures in MPL and use these structures to describe a number of elementary algorithms. The chapter includes a case study which describes an algorithm that solves a class of first order ordinary differential equations using the separation of variables technique and the method of exact equations with integrating factors.

Chapter 5: Recursive Algorithms. This chapter describes recursion as a programming technique in computer algebra and gives a number of examples that illustrate its advantages and limitations. It includes a case study that describes an elementary integration algorithm which finds the antiderivatives for a limited class of functions using the linear properties of the integral and the substitution method. Extensions of the algorithm to include the elementary rational function integration, some trigonometric integrals, elementary integration by parts, and one algebraic function form are described in the exercises.

Chapter 6: Structure of Polynomials and Rational Expressions. This chapter is concerned with the algorithms that analyze and manipulate polynomials and rational expressions. It includes computational definitions for various classes of polynomials and rational expressions that are based on the internal tree structure of expressions. Algorithms based on the primitive operations introduced in Chapter 3 are given for degree and coefficient computation, coefficient collection, expansion, and rationalization of algebraic expressions.

Chapter 7: Exponential and Trigonometric Transformations. This chapter is concerned with algorithms that manipulate exponential and trigonometric functions. It includes algorithms for exponential expansion and reduction, trigonometric expansion and reduction, and a simplification algorithm that can verify a large class of trigonometric identities.

Mathematical Methods

Chapter 1: Background Concepts. This chapter is a summary of the background material from *Elementary Algorithms* that provides a framework for the mathematical and computational discussions in the book. It includes a description of the mathematical psuedo-language (MPL), a brief discussion of the tree structure and polynomial structure of algebraic expressions, and a summary of the basic mathematical operators that appear in our algorithms.

Chapter 2: Integers, Rational Numbers, and Fields. This chapter is concerned with the numerical objects that arise in computer algebra, including integers, rational numbers, and algebraic numbers. It includes Euclid's algorithm for the greatest common divisor of two integers, the extended Euclidean algorithm, the Chinese remainder algorithm, and a simplification algorithm that transforms an involved arithmetic expression with integers and fractions to a rational number in standard form. In addition, it introduces the concept of a field which describes in a general way the properties of number systems that arise in computer algebra.

Chapter 3: Automatic Simplification. Automatic simplification is defined as the collection of algebraic and trigonometric simplification transformations that are applied to an expression as part of the evaluation process. In this chapter we take an in-depth look at the algebraic component of this process, give a precise definition of an automatically simplified expression, and describe an (involved) algorithm that transforms mathematical expressions to automatically simplified form. Although automatic simplification is essential for the operation of computer algebra software, this is the only detailed treatment of the topic in the textbook literature.

Chapter 4: Single Variable Polynomials. This chapter is concerned with algorithms for single variable polynomials with coefficients in a field. All algorithms in this chapter are ultimately based on polynomial division. It includes algorithms for polynomial division and expansion, Euclid's algorithm for greatest common divisor computation, the extended Euclidean algorithm, and a polynomial version of the Chinese remainder algorithm. In addition, the basic polynomial division and gcd algorithms are used to give algorithms for numerical computations in elementary algebraic number fields. These algorithms are then used to develop division and gcd algorithms for polynomials with algebraic number coefficients. The chapter concludes with an algorithm for partial fraction expansion that is based on the extended Euclidean algorithm.

Chapter 5: Polynomial Decomposition. Polynomial decomposition is a process that determines if a polynomial can be represented as a composition of lower degree polynomials. In this chapter we discuss some theoretical aspects of the decomposition problem and give an algorithm based on polynomial factorization that either finds a decomposition or determines that no decomposition exists.

Chapter 6: Multivariate Polynomials. This chapter generalizes the division and gcd algorithms to multivariate polynomials with coefficients in an integral domain. It includes algorithms for three polynomial division operations (recursive division, monomial-based division, and pseudo-division); polynomial expansion (including an application to the algebraic substitution problem); and the primitive and subresultant algorithms for gcd computation.

Chapter 7: The Resultant. This chapter introduces the resultant of two polynomials, which is defined as the determinant of a matrix whose entries depend on the coefficients of the polynomials. We describe a Euclidean algorithm and a subresultant algorithm for resultant computation and use the resultant to find polynomial relations for explicit algebraic numbers.

Chapter 8: Polynomial Simplification with Side Relations. This chapter includes an introduction to Gröbner basis computation with an application to the polynomial simplification problem. To simplify the presentation, we assume that polynomials have rational number coefficients and use the lexicographical ordering scheme for monomials.

Chapter 9: Polynomial Factorization. The goal of this chapter is the description of a basic version of a modern factorization algorithm for single variable polynomials in $\mathbf{Q}[x]$. It includes square-free factorization algorithms in $\mathbf{Q}[x]$ and $\mathbf{Z}_p[x]$, Kronecker's classical factorization algorithm for $\mathbf{Z}[x]$, Berlekamp's algorithm for factorization in $\mathbf{Z}_p[x]$, and a basic version of the Hensel lifting algorithm.

Computer Algebra Software and Programs

We use a procedure style of programming that corresponds most closely to the programming structures and style of the Maple, Mathematica, and MuPAD systems and, to a lesser degree, to the Macsyma and Reduce systems. In addition, some algorithms are described by transformation rules that translate to the pattern matching languages in the Mathematica and Maple systems. Unfortunately, the programming style used here does not translate easily to the structures in the Axiom system.

The dialogues and algorithms in these books have been implemented in the Maple 7.0, Mathematica 4.1, and MuPAD Pro (Version 2.0) systems. The dialogues and programs are found on a CD included with the books. In each book, available dialogues and programs are indicated by the word "Implementation" followed by a system name Maple, Mathematica, or MuPAD. System dialogues are in a notebook format (mws in Maple, nb in Mathematica, and mnb in MuPAD), and procedures are in text (ASCII) format (for examples, see the dialogue in Figure 1.1 on page 3 and the procedure in Figure 4.15 on page 148). In some examples, the dialogue display of a computer algebra system given in the text has been modified so that it fits on the printed page.

Electronic Version of the Book

These books have been processed in the LATEX 2_ε system with the *hyperref* package, which allows hypertext links to chapter numbers, section numbers, displayed (and numbered) formulas, theorems, examples, figures, footnotes, exercises, the table of contents, the index, the bibliography, and web sites. An electronic version of the book (as well as additional reference files) in the portable document format (PDF), which is displayed with the Adobe Acrobat software, is included on the CD.

Acknowledgements

I am grateful to the many students and colleagues who read and helped debug preliminary versions of this book. Their advice, encouragement, suggestions, criticisms, and corrections have greatly improved the style and structure of the book. Thanks to Norman Bleistein, Andrew Burt, Alex Champion, the late Jack Cohen, Robert Coombe, George Donovan, Bill Dorn, Richard Fateman, Clayton Ferner, Carl Gibbons, Herb Greenberg, Jillane Hutchings, Lan Lin, Zhiming Li, Gouping Liu, John Magruder, Jocelyn Marbeau, Stanly Steinberg, Joyce Stivers, Sandhya Vinjamuri, and Diane Wagner.

I am grateful to Gwen Diaz and Alex Champion for their help with the LATEX document preparation; Britta Wienand, who read most of the text and translated many of the programs to the MuPAD language; Aditya Nagrath, who created some of the figures; and Michael Wester who translated many of the programs to the Mathematica, MuPAD, and Macsyma languages. Thanks to Camie Bates, who read the entire manuscript and made numerous suggestions that improved the exposition and notation, and helped clarify confusing sections of the book. Her careful reading discovered numerous typographical, grammatical, and mathematical errors.

I also acknowledge the long-term support and encouragement of my home institution, the University of Denver. During the writing of the book, I was awarded two sabbatical leaves to develop this material.

Special thanks to my family for encouragement and support: my late parents Elbert and Judith Cohen, Daniel Cohen, Fannye Cohen, and Louis and Elizabeth Oberdorfer.

Finally, I would like to thank my wife, Kathryn, who as long as she can remember, has lived through draft after draft of this book, and who with patience, love, and support has helped make this book possible.

Joel S. Cohen
Denver, Colorado
September, 2001

1

Introduction to Computer Algebra

1.1 Computer Algebra and Computer Algebra Systems

The mathematical scientist models natural phenomena by translating experimental results and theoretical concepts into mathematical expressions containing numbers, variables, functions, and operators. Then, using accepted methods of mathematical reasoning, these expressions are carefully manipulated or transformed into other expressions that reveal new knowledge about the phenomenon being studied. This mathematical approach to understanding the world has been an important component of the scientific method in the physical sciences since the time of Galileo and Descartes. Following in the footsteps of these scientists, Isaac Newton used this approach to formulate an axiomatic, quantitative description of the motion of objects. By using mathematical reasoning, he discovered the universal law of gravitation and derived additional laws that describe the motion of the tides and the orbits of the planets. Thus the science we call *mechanics* was born, and the technique of manipulating and transforming mathematical expressions was firmly established as an important tool for discovering new knowledge about the physical world.

In the past fifty years, the computer has become an indispensable experimental tool that greatly extends our ability to solve mathematical problems. Mathematical scientists routinely use computers to obtain numerical and graphical solutions to problems that are too difficult or even impossible to solve by hand. But computers are not just number crunchers. In fact, at a basic level, computers simply manipulate symbols (0s and 1s) according to well-defined rules, and it is natural to ask what other parts of the math-

1

ematical reasoning process are amenable to computer implementation. Of course, it is unreasonable to expect a machine to formulate the axioms of mechanics as Newton did or derive from scratch the important results of the theory. However, one part of the mathematical reasoning process, the mechanical manipulation and analysis of mathematical expressions, is surprisingly algorithmic. There are now computer programs that routinely simplify algebraic expressions, integrate complicated functions, find exact solutions to differential equations, and perform many other operations encountered in applied mathematics, science, and engineering.

In this book we are concerned primarily with the development and application of algorithms and computer programs that carry out this mechanical aspect of the mathematical reasoning process. The field of mathematics and computer science that is concerned with this problem is known as *computer algebra* or *symbol manipulation*.

Computer Algebra Systems and Languages

A *computer algebra system* (CAS) or *symbol manipulation system* is a computer program that performs symbolic mathematical operations. In Figure 1.1 we show an interactive dialogue with the Maple computer algebra system developed by Waterloo Maple Inc. The statements that are preceded by the prompt (>) are inputs to the system that are entered at a computer workstation. The commands factor, convert, compoly, and simplify are examples of *mathematical operators* in the Maple system. In response to these statements, the program performs a mathematical operation and displays the result using a notation that is similar to ordinary mathematical notation.

In Figure 1.1, at the first two prompts, a polynomial is assigned (with the operator ":=") to a variable $u1$ and then factored in terms of irreducible factors with respect to the rational numbers. (In other words, none of the polynomials in the factored form can be factored further without introducing radicals.) At prompts three and four, we enter a rational expression and then find its partial fraction decomposition. At the next two prompts, Maple's compoly command determines that the polynomial $u3$ is a composite

$$u3 = f(g(x)), \quad f(x) = x^3 + 10 + 8\,x + 3\,x^2, \quad g(x) = 3x + x^2.$$

The process of representing a polynomial as a composite of lower degree polynomials is called *polynomial decomposition*. At the remaining prompts,

---

```
> u1 := x^5-4*x^4-7*x^3-41*x^2-4*x+35;
```

$$u1 := x^5 - 4x^4 - 7x^3 - 41x^2 - 4x + 35$$

```
> factor(u1);
```

$$(x+1)(x^2+2x+7)(x^2-7x+5)$$

```
> u2 := (x^4+7*x^2+3)/(x^5+x^3+x^2+1);
```

$$u2 := \frac{x^4 + 7x^2 + 3}{x^5 + x^3 + x^2 + 1}$$

```
> convert(u2,parfrac,x);
```

$$\frac{11}{6}\frac{1}{x+1} + \frac{\frac{2}{3}(4+x)}{x^2-x+1} - \frac{3}{2}\frac{x+1}{x^2+1}$$

```
> u3 := x^6+9*x^5+30*x^4+45*x^3+35*x^2+24*x+10;
```

$$u3 := x^6 + 9x^5 + 30x^4 + 45x^3 + 35x^2 + 24x + 10$$

```
> compoly(u3,x);
```

$$x^3 + 10 + 8x + 3x^2, \ x = 3x + x^2$$

```
> u4 := 1/(1/a+c/(a*b))+(a*b*c+a*c^2)/(b+c)^2;
```

$$u4 := \frac{1}{\dfrac{1}{a} + \dfrac{c}{ab}} + \frac{abc + ac^2}{(b+c)^2}$$

```
> simplify(u4);
```

$$a$$

```
> u5 := (sin(x)+sin(3*x)+sin(5*x)+sin(7*x))/(cos(x)+cos(3*x)
        +cos(5*x)+cos(7*x))-tan(4*x);
```

$$u5 := \frac{\sin(x) + \sin(3x) + \sin(5x) + \sin(7x)}{\cos(x) + \cos(3x) + \cos(5x) + \cos(7x)} - \tan(4x)$$

```
> simplify(u5);
```

$$0$$

Figure 1.1. An interactive dialogue with the Maple system that shows some symbolic operations from algebra and trigonometry. (Implementation: Maple (mws), Mathematica (nb), MuPAD (mnb).)

```
> u6 := cos(2*x+3)/(x^2+1);
```

$$u6 := \frac{\cos(2\,x+3)}{x^2+1}$$

```
> diff(u6,x);
```

$$-2\,\frac{\sin(2\,x+3)}{x^2+1} - 2\,\frac{\cos(2\,x+3)\,x}{(x^2+1)^2}$$

```
> u7 := cos(x)/(sin(x)^2+3*sin(x)+4);
```

$$u7 := \frac{\cos(x)}{\sin(x)^2 + 3\sin(x) + 4}$$

```
> int(u7,x);
```

$$\frac{2}{7}\,\sqrt{7}\arctan\left(\frac{1}{7}\,(2\sin(x)+3)\,\sqrt{7}\right)$$

```
> u8 := diff(y(x),x) + 3*y(x) = x^2+sin(x);
```

$$u8 := \left(\frac{\partial}{\partial x}\,\mathrm{y}(x)\right) + 3\,\mathrm{y}(x) = x^2 + \sin(x)$$

```
> dsolve(u8,y(x));
```

$$\mathrm{y}(x) = \frac{1}{3}\,x^2 - \frac{2}{9}\,x + \frac{2}{27} - \frac{1}{10}\,\cos(x) + \frac{3}{10}\,\sin(x) + e^{(-3\,x)}\,\_C1$$

Figure 1.2. An interactive dialogue with the Maple system that shows some symbolic operations from calculus and differential equations. (Implementation: Maple (mws), Mathematica (nb), MuPAD (mnb).)

Maple simplifies an involved algebraic expression $u4$ and then verifies a trigonometric identity[1].

In Figure 1.2, we again call on Maple to perform some operations from calculus and differential equations[2]. The `diff` command at the second

---

[1] Algebraic simplification is described in Sections 2.2 and 6.5, and trigonometric simplification is described in Section 7.2.

For further study, the reader may consult Cohen [24]: algebraic simplification is discussed in Chapter 3, Section 6.3, and Chapter 8; partial fraction decomposition in Section 4.4; polynomial decomposition in Chapter 5; and polynomial factorization in Chapter 9.

[2] We give algorithms in the book for all of these operations. Differentiation is described in Section 5.2, elementary integration in Section 5.3, and the solution of differential equations in Section 4.3.

prompt is used for differentiation and the `int` command at the fourth
prompt is for integration. Notice that the output of the `int` operator does
not include the arbitrary constant of integration. At the fifth prompt we
assign a first order differential equation[3] to `u7`, and at the sixth prompt
ask Maple to solve the differential equation. The symbol _C1 is Maple's
way of including an arbitrary constant in the solution[4].

We use the term *computer algebra language* or *symbolic programming
language* to refer to the computer language that is used to interact with a
CAS. Most computer algebra systems can operate in a *programming mode*
as well as an *interactive mode* (shown in Figures 1.1 and 1.2). In the programming mode, the mathematical operators `factor`, `simplify`, etc., are
combined with standard programming constructs such as assignment statements, loops, conditional statements, and subprograms to create programs
that solve more involved mathematical problems.

To illustrate this point, consider the problem of finding the formula for
the tangent line to the curve

$$y = f(x) = x^2 + 5x + 6$$

at the point $x = 2$. First, we find a general formula for the slope by
differentiation

$$\frac{dy}{dx} = 2x + 5.$$

The slope at the point $x = 2$ is obtained by substituting this value into
this expression

$$m = \frac{dy}{dx}(2) = 2(2) + 5 = 9.$$

The equation for the tangent line is obtained using the point slope form
for a line:

$$
\begin{aligned}
y = m(x - 2) + f(2) \ &= \ 9(x - 2) + 20 \qquad\qquad (1.1)\\
&= \ 9x + 2.
\end{aligned}
$$

To obtain the last formula, we have expanded the right side of Equation (1.1).

In Figure 1.3 we give a general procedure, written in the Maple computer algebra language that mimics these calculations. The procedure computes the tangent line formula for an arbitrary expression f at the point

---

[3]Maple displays the derivative of an unknown function `y(x)` using the partial derivative symbol instead of ordinary derivative notation.

[4]Maple includes an arbitrary constant in the solution of a differential equation, but
does not include the arbitrary constant for an antidifferentiation. Inconsistencies of this
sort are commonplace with computer algebra software.

```
1  Tangent_line := proc(f,x,a)
2  local
3    deriv,m,line;
4  deriv := diff(f,x);
5  m := subs(x=a,deriv);
6  line := expand(m*(x-a)+subs(x=a,f));
7  RETURN(line)
8  end:
```

Figure 1.3. A procedure in the Maple language that obtains a formula for the tangent line. The line numbers are not part of the Maple program. (Implementation: Maple (txt), Mathematica (txt), MuPAD (txt).)

$x = a$. The operator diff in line 4 is used for differentiation and the operator subs in line 5 for substitution. The expand operator in line 6 is included to simplify the output. Once the procedure is entered into the Maple system, it can be invoked from the interactive mode of the system (see Figure 1.4).

```
> Tangent_line(x^2+5*x+6, x, 2);
```

$$9x + 2$$

Figure 1.4. The execution of the Tangent_line procedure in the interactive mode of the Maple system. (Implementation: Maple (mws), Mathematica (nb), MuPAD (mnb).)

Commercial Computer Algebra Systems

In the last 15 years, we have seen the creation and widespread distribution of a number of large (but easy to use) computer algebra systems. The most prominent of the commercial and University packages are:

- **Axiom** – a very large CAS originally developed at IBM under the name Scratchpad. Information about Axiom can be found in Jenks and Sutor [50].

- **Derive** – a small CAS originally designed by Soft Warehouse Inc. for use on a personal computer. Derive has also been incorporated in the

TI-89 and TI-92 handheld calculators produced by Texas Instruments Inc. Information about Derive can be found at the web site

> http://www.derive.com.

- **Macsyma** – a very large CAS originally developed at M.I.T. in the late 1960s and 1970s. There are currently a number of versions of the original Macsyma system. Information about Macsyma can be found in Wester [100].

- **Maple** – a very large CAS originally developed by the Symbolic Computation Group at the University of Waterloo (Canada) and now distributed by Waterloo Maple Inc. Information about Maple is found in Heck [45] or at the web site

> http://www.maplesoft.com.

- **Mathematica** – a very large CAS developed by Wolfram Research Inc. Information about Mathematica can be found in Wolfram [102] or at the web site

> http://www.wolfram.com.

- **MuPAD** – a large CAS developed by the University of Paderborn (Germany) and SciFace Software GmbH & Co. KG. Information about MuPAD can be found in Gerhard et al. [40] or at the web site

> http://www.mupad.com.

- **Reduce** – one of the earliest computer algebra systems originally developed in the late 1960s and 1970s. Information about Reduce is found in Rayna [83] or at the web site

> http://www.uni-koeln.de/REDUCE.

All of these packages are integrated mathematics problem solving systems that include facilities for exact symbolic computations (similar to those in Figures 1.1, 1.2, and 1.3), along with some capability for (approximate) numerical solution of mathematical problems and high quality graphics. The examples in this book refer primarily to the computer algebra capabilities of the Maple, Mathematica, and MuPAD systems, since these systems are readily available and support a programming style that is most similar to the one used here.

Mathematical Knowledge in Computer Algebra Systems

Computer algebra systems have the capability to perform exact symbolic computations in many areas of mathematics. A sampling of these capabilities includes:

- **Arithmetic** – unlimited precision rational number arithmetic, complex (rational number) arithmetic, transformation of number bases, interval arithmetic, modulo arithmetic, integer operations (greatest common divisors, least common multiples, prime factorization), combinatorial functions.

- **Algebraic manipulation** – simplification, expansion, factorization, substitution operations.

- **Polynomial operations** – structural operations on polynomials (degree, coefficient extraction), polynomial division, greatest common divisors, factorization, resultant calculations, polynomial decomposition, simplification with respect to side relations.

- **Solution of equations** – polynomial equations, some non-linear equations, systems of linear equations, systems of polynomial equations, recurrence relations.

- **Trigonometry** – trigonometric expansion and reduction, verification of identities.

- **Calculus** – derivatives, antiderivatives, definite integrals, limits, Taylor series, manipulation of power series, summation of series, operations with the special functions of mathematical physics.

- **Differential equations** – solution of ordinary differential equations, solution of systems of differential equations, solution using series, solution using Laplace transforms, solution of some partial differential equations.

- **Advanced algebra** – manipulations with algebraic numbers, group theory, Galois groups.

- **Linear algebra and related topics** – matrix operations, vector and tensor analysis.

- **Code generation** – formula translation to conventional programming languages such as FORTRAN and C, formula translation to mathematics word processing languages (LaTeX).

In addition, computer algebra systems have the capability to utilize this mathematical knowledge in computer programs that solve other mathematical problems.

Exercises

1. What transformation rules from algebra, trigonometry, or calculus must a computer "know" to perform the following operations? Be careful not to omit any obvious arithmetic or algebraic rules that are used to obtain the result in a simplified form.

   (a) $\dfrac{d(a\,x + x\,e^{x^2})}{dx} = a + e^{x^2} + 2x^2 e^{x^2}.$

   (b) $\dfrac{\sec(x)}{\sin(x)} - \dfrac{\sin(x)}{\cos(x)} - \cot(x) = 0.$

   (c) $\dfrac{1}{1/a + c/(a\,b)} + \dfrac{a\,b\,c + a\,c^2}{(b+c)^2} = a.$

2. All computer algebra systems include an algebraic expansion command that obtains transformations similar to

$$
\begin{aligned}
(x+2)(x+3)(x+4) &= x^3 + 9x^2 + 26x + 24, \\
(x+y+z)^3 &= x^3 + y^3 + z^3 + 3x^2 y + 3x^2 z + 3y^2 x \\
&\quad + 3y^2 z + 3z^2 x + 3z^2 y + 6x\,y\,z, \\
(x+1)^2 + (y+1)^2 &= x^2 + 2x + y^2 + 2y + 2, \\
((x+2)^2 + 3)^2 &= x^4 + 8x^3 + 30x^2 + 56x + 49.
\end{aligned}
$$

   (In Maple, the **expand** command; in Mathematica the **Expand** command; in MuPAD, the **expand** command.)

   What algorithm would you use to perform this operation? It is not necessary to give the exact algorithm. Rather describe some of the issues that arise when you try to design a mechanical procedure for this operation. What mathematical and computational techniques are useful for this algorithm?

3. The simplification of mathematical expressions is an important aspect of the mathematical reasoning process and all computer algebra systems have some capability to perform this operation (see Figure 1.1 on page 3). Although simplification is described in elementary mathematics textbooks, it is defined in a vague way. However, to give an algorithm that performs simplification, we must have a precise definition of the term. Is it possible to give a precise definition for simplification?

1.2   Applications of Computer Algebra

The Purpose of Applied Mathematics

In the fascinating book *Mathematics Applied to Deterministic Problems in the Natural Sciences* ([63], SIAM, 1988, pages 5-7), Lin and Segel describe the purpose of applied mathematics in the following way:

> The purpose of applied mathematics is to elucidate scientific concepts and describe scientific phenomena through the use of mathematics, and to stimulate the development of new mathematics through such studies.

They discuss three aspects of this process that relate to the solution of scientific problems:

(i) the **formulation** of the scientific problem in mathematical terms.

(ii) the **solution** of the mathematical problems thus created.

(iii) the **interpretation** of the solution and its empirical verification in scientific terms.

In addition, they mention a closely related adjunct of this process:

(iv) the generation of scientifically relevant new mathematics through creation, generalization, abstraction, and axiomatic formulation.

In principle, computer algebra can help facilitate steps (i), (ii), and (iv) of this process. In practice, computer algebra is primarily involved in step (ii) and to a much lesser degree in steps (i) and (iv).

Examples of Computer Algebra

In the remainder of this section, we give four examples that illustrate the use of computer algebra software in the problem solving process. All of the examples are concerned with the solution of equations.

Example 1.1.   (Solution of a linear system of equations.)   A CAS is particularly useful for calculations that are lengthy and tedious but straightforward. The solution of a linear system of equations with symbolic coefficients provides an example of this situation. The following system of equations occurs in a problem in statistical mechanics[5]:

---

[5] The author encountered this system of equations while working on a problem in statistical mechanics in 1982. At that time the solution of the system with pencil and paper (including checking and re-checking the result) took two days. Unfortunately, the published result still contains a minor coefficient error. See Cohen, Haskins, and Marchand [23].

$$\begin{aligned}
d_0 + d_1 + d_2 + d_3 + d_4 &= 1, \\
d_1 + 2d_2 + 3d_3 + 4d_4 &= 2(1 - m), \\
3d_0 - d_2 + 3d_4 &= 2\gamma_{2,0} + \gamma_{1,1}, \\
\phi d_0 + \psi d_1 - \psi d_3 - \phi d_4 &= m, \\
2\phi d_0 + \psi d_1 + \psi d_3 + 2\phi d_4 &= 2\gamma_{1,0}.
\end{aligned} \tag{1.2}$$

In this system the five unknown variables are $d_0, d_1, d_2, d_3,$ and d_4. The coefficients of these variables and the right-hand sides of the equations depend on the six parameters $m, \phi, \psi, \gamma_{1,0}, \gamma_{1,1},$ and $\gamma_{2,0}$, and the object is to express the unknowns in terms of these parameters. Whether or not this is a good problem for a CAS depends on the purpose of the computation. In this case a solution is needed to help understand the effect of the various parameters on the individual unknowns. What is needed is not just a solution, but one that is compact enough to allow for an easy interpretation of the result.

The symbolic solution of five linear equations with five unknowns has the potential to produce expressions with hundreds of terms. In this case, however, the coefficients are not completely random but instead contain a symmetry pattern. Because of this there is reason to believe (but no guarantee) that the solutions will simplify to expressions of reasonable size.

Figure 1.5 shows an interactive dialogue with the Mathematica system that solves the system of equations. The input statements in Mathematica are indicated by the label "*In*" followed by an integer in brackets and the symbol "*:=*" (*In[1]:=, In[2]:=,* etc.). The symbols *Out[1]=, Out[2]=,* etc., are labels that represent the output produced by each input line. The other equal sign in lines *In[1]* through *In[6]* is an assignment symbol and the symbol "*==*" is used for equality in an equation. The command to solve the system of equations is given in *In[6]* and the solution to the system is displayed in the lines following *Out[6]*. As we suspected, the solution simplifies to expressions of reasonable size. □

One application of computer algebra systems is the exact solution of polynomial equations. For polynomial equations with degree less than or equal to four it is always possible to obtain solutions in terms of expressions with radicals, although for cubic and quartic equations these solutions are often quite involved. For polynomials with degree five or greater, it is theoretically impossible to represent the solutions of all such equations using expressions with radicals[6], although it is possible to solve some of these equations.

---

[6]This statement follows from Galois theory, the algebraic theory that describes the nature of solutions to polynomial equations.

$In[1] :=$ eq1 = d[0] + d[1] + d[2] + d[3] + d[4] == 1

$Out[1] = d[0] + d[1] + d[2] + d[3] + d[4] == 1$

$In[2] :=$ eq2 = d[1] + 2 * d[2] + 3 * d[3] + 4 * d[4] == 2 * (1 - m)

$Out[2] = d[1] + 2d[2] + 3d[3] + 4d[4] == 2(1 - m)$

$In[3] :=$ eq3 = 3 * d[0] - d[2] + 3 * d[4] == 2 * $\gamma[2, 0] + \gamma[1, 1]$

$Out[3] := 3d[0] - d[2] + 3d[4] == \gamma[1, 1] + 2\gamma[2, 0]$

$In[4] :=$ eq4 = ϕ * d[0] + φ * d[1] - φ * d[3] - ϕ * d[4] == m

$Out[4] := \phi d[0] + \varphi d[1] - \varphi d[3] - \phi d[4] == m$

$In[5] :=$ eq5 = 2 * ϕ * d[0] + φ * d[1] + φ * d[3] + 2 * ϕ * d[4] == 2 * $\gamma[1, 0]$

$Out[5] := 2\phi d[0] + \varphi d[1] + \varphi d[3] + 2\phi d[4] == 2\gamma[1, 0]$

$In[6] :=$ Solve[{eq1, eq2, eq3, eq4, eq5}, {d[0], d[1], d[2], d[3], d[4]}]

$$Out[6] = \left\{ \left\{ d[2] \to -\frac{1}{2(\phi - 2\varphi)}(3\varphi - 6\gamma[1, 0] + 2\phi\gamma[1, 1] - \varphi\gamma[1, 1] + 4\phi\gamma[2, 0] \right. \right.$$
$$-2\varphi\gamma[2, 0]),$$
$$d[0] \to -\frac{1}{4(\phi - 2\varphi)}(-2m + \varphi + 4m\varphi - 2\gamma[1, 0] + \varphi\gamma[1, 1] + 2\varphi\gamma[2, 0]),$$
$$d[1] \to -\frac{1}{2(\phi - 2\varphi)}(2m - \phi - 2m\phi + 4\gamma[1, 0] - \phi\gamma[1, 1] - 2\phi\gamma[2, 0]),$$
$$d[3] \to -\frac{1}{2(\phi - 2\varphi)}(-2m - \phi + 2m\phi + 4\gamma[1, 0] - \phi\gamma[1, 1] - 2\phi\gamma[2, 0]),$$
$$\left. \left. d[4] \to -\frac{1}{4(\phi - 2\varphi)}(2m + \varphi - 4m\varphi - 2\gamma[1, 0] + \varphi\gamma[1, 1] + 2\varphi\gamma[2, 0]) \right\} \right\}$$

Figure 1.5. An interactive dialogue with the Mathematica system that solves a system of linear equations. (Implementation: Maple (mws), Mathematica (nb), MuPAD (mnb).)

Example 1.2. (**Solution of cubic polynomial equations.**) To examine the possibilities (and limitations) for symbolic solutions of polynomial equations, consider the cubic equation

$$x^3 - 2ax + a^3 = 0 \tag{1.3}$$

where the symbol a is a parameter. We examine the nature of the solution for various values of a using the Maple system[7] in Figures 1.6, 1.7, and 1.8. At the first prompt (>) in Figure 1.6, the equation is assigned to the variable eq. At the second prompt, the equation is solved for x using Maple's solve command and stored in the variable general_solution. The involved solution, which contains three expressions separated by commas, is expressed in terms of an auxiliary expression for which Maple has chosen the name %1 and the symbol I which represents $\sqrt{-1}$. In ordinary (and more user friendly) mathematical notation the three solutions are

$$
x = \frac{1}{6}r^{1/3} + \frac{4\,a}{r^{1/3}},
$$

$$
-\frac{1}{12}r^{1/3} - \frac{2\,a}{r^{1/3}} + 1/2\,\imath\,\sqrt{3}\left(\frac{1}{6}r^{1/3} - \frac{4\,a}{r^{1/3}}\right),
$$

$$
-\frac{1}{12}r^{1/3} - \frac{2\,a}{r^{1/3}} - 1/2\,\imath\,\sqrt{3}\left(\frac{1}{6}r^{1/3} - \frac{4\,a}{r^{1/3}}\right),
$$

where

$$
r = -108\,a^3 + 12\,\sqrt{-96\,a^3 + 81\,a^6}, \qquad \imath = \sqrt{-1}.
$$

At the next prompt the subs command[8] is used to substitute $a = 1$ in the general solution to obtain the solution $s1$. In this form the expressions are so involved that it is difficult to tell which roots are real numbers and which ones have an imaginary part. Since a cubic equation with real number coefficients can have at most two roots with non-zero imaginary parts, at least one of the roots must be a real number. At the fourth prompt, we attempt to simplify the solutions with Maple's radsimp command, which can simplify some expressions with radicals[9]. In this case, unfortunately, it only transforms the solution to another involved form.

To determine the nature (real or not real) of the roots, at the next prompt we apply Maple's evalc command, which expresses the roots in

---

[7]For the Maple dialogues in this section, the Output Display is set to the Typeset Notation option. Other options display output expressions in other forms.

[8]This input statement has one unfortunate complication. Observe that in the subs command we have placed the set braces { and } about general_solution. The reason for this has to do with the form of the output of Maple's solve command. For this equation, general_solution consists of three expressions separated by commas which is known as an *expression sequence* in the Maple language. Unfortunately, an expression sequence cannot be input for the Maple subs command, and so we have included the two braces so that the input expression is now a Maple *set* which is a valid input. Observe that the output $s1$ is also a Maple set.

[9]Another possibility is the Maple command radsimp(s1,ratdenom) which is an optional form that rationalizes denominators. This command obtains a slightly different form, but not the simplified form.

```
> eq := x^3-2*a*x+a^3=0;
```

$$eq := x^3 - 2\,a\,x + a^3 = 0$$

```
> general_solution := solve(eq,x);
```

$$general\_solution := \frac{1}{6}\,\%1^{(1/3)} + \frac{4\,a}{\%1^{(1/3)}},$$

$$-\frac{1}{12}\,\%1^{(1/3)} - \frac{2\,a}{\%1^{(1/3)}} + \frac{1}{2}\,I\,\sqrt{3}\,(\frac{1}{6}\,\%1^{(1/3)} - \frac{4\,a}{\%1^{(1/3)}}),$$

$$-\frac{1}{12}\,\%1^{(1/3)} - \frac{2\,a}{\%1^{(1/3)}} - \frac{1}{2}\,I\,\sqrt{3}\,(\frac{1}{6}\,\%1^{(1/3)} - \frac{4\,a}{\%1^{(1/3)}})$$

$$\%1 := -108\,a^3 + 12\,\sqrt{-96\,a^3 + 81\,a^6}$$

```
> s1 := subs(a=1,{general_solution});
```

$$s1 := \left\{ \frac{1}{6}\,\%1 + \frac{4}{\left(-108 + 12\,\sqrt{-15}\right)^{1/3}}, \right.$$

$$-\frac{1}{12}\,\%1 - \frac{2}{\left(-108 + 12\,\sqrt{-15}\right)^{1/3}} + \frac{1}{2}\,I\,\sqrt{3}\left(\frac{1}{6}\,\%1 - \frac{1}{\left(-108 + 12\,\sqrt{-15}\right)^{1/3}} \right),$$

$$\left. -\frac{1}{12}\,\%1 - \frac{2}{\left(-108 + 12\,\sqrt{-15}\right)^{1/3}} - \frac{1}{2}\,I\,\sqrt{3}\left(\frac{1}{6}\,\%1 - \frac{1}{\left(-108 + 12\,\sqrt{-15}\right)^{1/3}} \right) \right\}$$

$$\%1 := \left(-108 + 12\,\sqrt{-15}\right)^{1/3}$$

```
> radsimp(s1);
```

$$\left\{ \frac{1}{6}\,\frac{\left(-108 + 12\,I\sqrt{15}\right)^{(2/3)} + 24}{\sqrt[3]{-108 + 12\,I\sqrt{15}}},\; \frac{1}{12}\,\frac{-\%1^{(2/3)} - 24 + I\sqrt{3}\,\%1^{(2/3)} - 24\,I\sqrt{3}}{\%1}, \right.$$

$$\left. -\frac{1}{12}\,\frac{\%1^{(2/3)} + 24 + I\sqrt{3}\,\%1^{(2/3)} - 24\,I\sqrt{3}}{\%1}, \right\}$$

$$\%1 := -108 + 12\,I\,\sqrt{3}\,\sqrt{5}$$

```
> simplify(evalc(s1))
```

$$\left\{ -\frac{1}{3}\,\sqrt{2}(\sqrt{3}\cos(\%1) + 3\sin(\%1)), -\frac{1}{3}\,\sqrt{2}(\sqrt{3}\cos(\%1) - 3\sin(\%1)), \frac{2}{3}\sqrt{2}\,\sqrt{3}\cos(\%1) \right\}$$

$$\%1 := -\frac{1}{3}\arctan\left(\frac{1}{9}\sqrt{3}\,\sqrt{5} \right) + \frac{1}{3}\pi$$

Figure 1.6. An interactive dialogue with the Maple system for solving a cubic equation. (Implementation: Maple (mws), Mathematica (nb), MuPAD (mnb))

```
> eq2 := subs(a=1,eq);
```
$$eq2 := x^3 - 2x + 1 = 0$$
```
> solve(eq2,x);
```
$$1, \ -\frac{1}{2} + \frac{1}{2}\sqrt{5}, \ -\frac{1}{2} - \frac{1}{2}\sqrt{5}$$

Figure 1.7. Solving a cubic equation with Maple (continued). (Implementation: Maple (mws), Mathematica (nb), MuPAD (mnb).)

terms of their real and imaginary parts, and then apply the simplify command, which attempts to simplify the result. Observe that the solutions are now expressed in terms of the trigonometric functions sin and cos and the inverse function arctan. Although the solutions are still quite involved, we see that all three roots are real numbers. We will show below that the solutions can be transformed to a much simpler form, although this cannot be done directly with these Maple commands.

Actually, a better approach to find the roots when $a = 1$ is to substitute this value in Equation (1.3), and solve this particular equation rather than use the general solution. This approach is illustrated in Figure 1.7. At the first prompt we define a new equation $eq2$, and at the second prompt solve the equation. In this case the roots are much simpler since Maple can factor the polynomial as $x^3 - 2x + 1 = (x - 1)(x^2 + x - 1)$ which leads to simple exact expressions. On the other hand the general equation (1.3) cannot be factored for all values of a, and so the roots in Figure 1.6 for $a = 1$ are given by much more involved expressions.

This example illustrates an important maxim about computer algebra:

A general approach to a problem should be avoided when a particular solution will suffice.

Although the general solution gives a solution for $a = 1$, the expressions are unnecessarily involved, and to obtain useful information requires an involved simplification, which cannot be done easily with the Maple software[10].

Let's consider next the solution of Equation (1.3) when $a = 1/2$. In Figure 1.8, at the first prompt we define a new cubic equation $eq3$, and

---

[10]This simplification can be done with the Mathematica system using the FullSimplify command and with the MuPAD system using the radsimp command. There are, however, other examples that cannot be simplified by any of the systems. See Footnote 6 on page 145 for a statement about the theoretical limitations of algorithmic simplification.

```
> eq3 := subs(a=1/2,eq);
```

$$eq3 := x^3 - x + \frac{1}{8} = 0$$

```
> s2:=solve(eq3,x);
```

$$s2 := \frac{1}{12}\%1 + \frac{4}{(-108 + 12\,I\,\sqrt{687})^{(1/3)}},$$

$$-\frac{1}{24}\%1 - \frac{2}{(-108 + 12\,I\,\sqrt{687})^{(1/3)}} + \frac{1}{4}I\sqrt{3}\left(\frac{1}{6}\%1 - \frac{8}{(-108 + 12\,I\,\sqrt{687})^{(1/3)}}\right),$$

$$-\frac{1}{24}\%1 - \frac{2}{(-108 + 12\,I\,\sqrt{687})^{(1/3)}} - \frac{1}{4}I\sqrt{3}\left(\frac{1}{6}\%1 - \frac{8}{(-108 + 12\,I\,\sqrt{687})^{(1/3)}}\right)$$

$$\%1 := (-108 + 12\,I\,\sqrt{687})^{(1/3)}$$

```
> s3 := radsimp({s2});
```

$$s3 := \left\{ \frac{1}{12}\frac{(-108 + 12\,I\sqrt{687})^{(2/3)} + 48}{(-108 + 12\,I\sqrt{687})^{(1/3)}},\ \frac{1}{24}\frac{-\%1^{(2/3)} - 48 + I\sqrt{3}\%1^{(2/3)} - 48\,I\sqrt{3}}{\%1^{(1/3)}}, \right.$$

$$\left. -\frac{1}{24}\frac{\%1^{(2/3)} + 48 + I\sqrt{3}\,\%1^{(2/3)} - 48\,I\sqrt{3}}{\%1^{(1/3)}} \right\}$$

$$\%1 := -108 + 12\,I\,\sqrt{3}\,\sqrt{229}$$

```
> simplify(evalc({s2}))
```

$$\left\{ \frac{2}{3}\sqrt{3}\cos(\%1),\ -\frac{1}{3}\sqrt{3}\cos(\%1) - \sin(\%1),\ -\frac{1}{3}\sqrt{3}\cos(\%1) + \sin(\%1) \right\}$$

$$\%1 := -\frac{1}{3}\arctan\left(\frac{1}{9}\sqrt{3}\,\sqrt{229}\right) + \frac{1}{3}\pi$$

```
> evalf(s3)
```

$$\{.9304029266 - .8624347141\,10^{-10}\,I,\ -1.057453771 + .4629268900\,10^{-9}\,I,$$
$$.1270508443 - .2120100566\,10^{-9}\,I\}$$

Figure 1.8. Solving cubic equations with Maple (continued). (Implementation: Maple (mws), Mathematica (nb), MuPAD (mnb).)

at the next three prompts solve it and try to simplify the roots. Again the representations of the roots in *s2* and *s3* are quite involved, and it is difficult to tell whether the roots are real or include imaginary parts. Again, to determine the nature of the roots, we apply Maple's `evalc` and `simplify` commands and obtain an involved representation in terms of the trigonometric functions sin and cos and the inverse function arctan. Although the solutions are still quite involved, it appears that all three roots are real numbers.

In this case, nothing can be done to simplify the exact roots. In fact, even though the three roots are real numbers, we can't eliminate the symbol $\imath = \sqrt{-1}$ from $s2$ or $s3$ without introducing the trigonometric functions as in $s4$. This situation, which occurs when none of the roots of a cubic equation is a rational number[11], shows that there is a theoretical limitation to how useful the exact solutions using radicals can be. The exact solutions can be found, but cannot be simplified to a more useful form.

Given this situation, at the last prompt, we apply Maple's `evalf` command that evaluates the roots $s3$ to an approximate decimal format. The small non-zero imaginary parts that appear in the roots are due to the round-off error that is inevitable with approximate numerical calculations.   □

Example 1.3. (**Solution of higher degree polynomial equations.**) Although computer algebra systems can solve some higher degree polynomial equations, they cannot solve all such equations, and in cases where solutions can be found they are often so involved that they are not useful in practice (Exercise 2(a)). Nevertheless, computer algebra systems can obtain useful solutions to some higher degree equations. This is shown in the first two examples in the MuPAD dialogue in Figure 1.9.

At the first prompt (the symbol •) we assign a polynomial to the variable u, and then at the next prompt solve $u = 0$ for x. In this case MuPAD obtains the solutions by first factoring u in terms of polynomials with integer coefficients as

$$u = (x - 1)\,(x^2 + x + 2)\,(x^2 + 5\,x - 4),$$

and then using the quadratic formula for the two quadratic factors.

At the third prompt we assign a sixth degree polynomial to v, and try to factor it at the next prompt. Since MuPAD returns the same polynomial, it is not possible to factor v in terms of lower degree polynomials that have integer coefficients. At the next prompt, however, MuPAD obtains the six roots to $v = 0$. In this case MuPAD finds the solutions by first recognizing that the polynomial v can be written as a composition of polynomials

$$v = f(g(x)), \quad f(w) = w^3 - 2, \quad w = g(x) = x^2 - 2\,x - 1.$$

---

[11] See Birkhoff and Mac Lane [10], page 450, Theorem 22. An interesting historical discussion of this problem is given in Nahin [74].

- u := x ∧ 5 + 5 ∗ x ∧ 4 − 3 ∗ x ∧ 3 + 3 ∗ x ∧ 2 − 14 ∗ x + 8;
  $$-14 \cdot x + 3 \cdot x^2 - 3 \cdot x^3 + 5 \cdot x^4 + x^5 + 8$$

- solve(u = 0, x, MaxDegree = 5);
  $$\left\{ 1, -\frac{\sqrt{41}}{2} - \frac{5}{2}, \frac{\sqrt{41}}{2} - \frac{5}{2}, \left(-\frac{\imath}{2}\right) \cdot \sqrt{7} - \frac{1}{2}, \left(\frac{\imath}{2}\right) \cdot \sqrt{7} - \frac{1}{2} \right\}$$

- v := x ∧ 6 − 6 ∗ x ∧ 5 + 4 ∗ x ∧ 3 + 9 ∗ x ∧ 4 − 9 ∗ x ∧ 2 − 6 ∗ x − 3;
  $$-6 \cdot x - 9 \cdot x^2 + 4 \cdot x^3 + 9 \cdot x^4 - 6 \cdot x^5 + x^6 - 3$$

- factor(v);
  $$-6 \cdot x - 9 \cdot x^2 + 4 \cdot x^3 + 9 \cdot x^4 - 6 \cdot x^5 + x^6 - 3$$

- solve(v = 0, x, MaxDegree = 6);
  $$\left\{ \sqrt{\sqrt[3]{2} + 2} + 1, -\sqrt{\sqrt[3]{2} + 2} + 1, -\frac{\sqrt{-8 \cdot \sqrt[3]{2} + (-8 \cdot \imath) \cdot \sqrt[3]{2} \cdot \sqrt{3} + 32}}{4} + 1, \right.$$
  $$\frac{\sqrt{-8 \cdot \sqrt[3]{2} + (-8 \cdot \imath) \cdot \sqrt[3]{2} \cdot \sqrt{3} + 32}}{4} + 1, \frac{-\sqrt{-8 \cdot \sqrt[3]{2} + (8 \cdot \imath) \cdot \sqrt[3]{2} \cdot \sqrt{3} + 32}}{4} + 1,$$
  $$\left. \frac{\sqrt{-8 \cdot \sqrt[3]{2} + (8 \cdot \imath) \cdot \sqrt[3]{2} \cdot \sqrt{3} + 32}}{4} + 1 \right\}$$

- w := x ∧ 8 − 136 ∗ x ∧ 7 + 6476 ∗ x ∧ 6 − 141912 ∗ x ∧ 5 + 1513334 ∗ x ∧ 4
  − 7453176 ∗ x ∧ 3 + 13950764 ∗ x ∧ 2 − 5596840 ∗ x + 46225;
  $$-5596840 \cdot x + 13950764 \cdot x^2 - 7453176 \cdot x^3 + 1513334 \cdot x^4 - 141912 \cdot x^5$$
  $$+ 6476 \cdot x^6 - 136 \cdot x^7 + x^8 + 46225$$

- solve(w = 0, x, MaxDegree = 8);
  $$RootOf \left(-5596840 \cdot X1 + 13950764 \cdot X1^2 - 7453176 \cdot X1^3 + 1513334 \cdot X1^4 \right.$$
  $$\left. -141912 \cdot X1^5 + 6476 \cdot X1^6 - 136 \cdot X1^7 + X1^8 + 46225, X1 \right)$$

- r := (sqrt(2) + sqrt(3) + sqrt(5) + sqrt(7)) ∧ 2;
  $$\left(\sqrt{2} + \sqrt{3} + \sqrt{5} + \sqrt{7} \right)^2$$

- expand(subs(w, x = r));
  $$0$$

Figure 1.9. The solution of high degree polynomial equations using MuPAD. (Implementation: Maple (mws), Mathematica (nb), MuPAD (mnb).)

In this form the solution to $v = 0$ is obtained by solving $w^3 - 2 = 0$ to obtain

$$w = 2^{1/3}, \quad -\frac{2^{1/3}}{2} + \frac{2^{1/3}3^{1/2}}{2} \, \imath, \quad -\frac{2^{1/3}}{2} - \frac{2^{1/3}3^{1/2}}{2} \, \imath,$$

and then solving the three equations

$$x^2 - 2x - 1 = 2^{1/3},$$

$$x^2 - 2x - 1 = -\frac{2^{1/3}}{2} + \frac{2^{1/3}\,3^{1/2}}{2}\,\imath,$$

$$x^2 - 2x - 1 = -\frac{2^{1/3}}{2} - \frac{2^{1/3}\,3^{1/2}}{2}\,\imath.$$

For example, by solving the first of these equations we obtain the first two roots of $v = 0$ in Figure 1.9.

Next, we assign an involved eighth degree polynomial to w, and attempt to solve the equation $w = 0$. Even though the equation has the eight roots

$$x = (\sqrt{2} \pm \sqrt{3} \pm \sqrt{5} \pm \sqrt{7})^2, \tag{1.4}$$

the MuPAD `solve` command is unable to find them, and returns instead a curious expression that simply says the solutions are roots of the original equation. At the next two prompts we assign to the variable r one of the roots in Equation (1.4), and then use the `subs` and `expand` commands to verify that it is a solution to the equation. $\square$

A Word of Caution

It goes without saying (but let's say it anyway), that there is more to mathematical reasoning than the mechanical manipulation of symbols. It is easy to give examples where a mechanical approach to mathematical manipulation leads to an incorrect result. This point is illustrated in the next example.

Example 1.4. Consider the following equation for x:

$$\sqrt{x + 7} + \sqrt{x + 2} = 1, \tag{1.5}$$

where we assume that the square root symbol represents a non-negative number and $x \geq -2$ so that the expressions under the radical signs are non-negative. Suppose that the goal is to find all real values of x that satisfy this equation. First transform the equation to

$$\sqrt{x + 7} = 1 - \sqrt{x + 2}. \tag{1.6}$$

Squaring both sides of this equation and simplifying gives

$$-2 = \sqrt{x + 2}. \tag{1.7}$$

By squaring both sides of this equation and solving for x, we obtain

$$x = 2. \tag{1.8}$$

However, this value is not a root of the original Equation (1.5). What is wrong with our reasoning?

In this case, the problem lies with the interpretation of the square root symbol. If we insist that the square roots are always non-negative, there are no real roots. However, if we allow (somewhat arbitrarily) the second square root in Equation (1.5) to be negative, the value $x = 2$ is a root. Indeed, the necessity of this assumption appears during the calculation in Equation (1.7).

Let's see what happens when we try to solve Equation (1.5) with a computer algebra system. Consider the dialogue with the Macsyma system in Figure 1.10. The input statements in Macsyma are preceded by the letter c followed by a positive integer ((c1), (c2), etc.). The symbols ((d1), (d2), etc.) are labels that represent the output produced by each input line. The colon in line (c1) is the assignment symbol in Macsyma. At line (c1), we assign the equation to the variable $eq1$ and at (c2) attempt to solve the equation for x. Observe that Macsyma simply returns the equation in a modified form indicating that it cannot solve the equation with its `solve` command.

We can, however, help Macsyma along by directing it to perform manipulations similar to the ones in Equations (1.6) through (1.8). At (c3), (c4), (c5), and (c6) we direct the system to put the equation in a form that can be solved for x at (c7). Again we obtain the extraneous root $x = 2$. Of course, at (c8) when we substitute this value into the original equation, we obtain an inequality since Macsyma assumes that all square roots of positive integers are positive.

This example shows that it is just as important to scrutinize our computer calculations as our pencil and paper calculations. The point is mathematical symbols have meaning, and transformations that are correct in one context may require subtle assumptions in other contexts that render them meaningless. In this simple example it is easy to spot the flaw in our reasoning. In a more involved example with many steps and involved output we may not be so lucky. Additional examples of how incorrect conclusions can follow from deceptive symbol manipulation are given in Exercises 10, 11, 12, and 13.   □

Exploring the Capabilities of a CAS

An important prerequisite for successful use of a CAS is an understanding of its capabilities and limitations. Since some symbolic operations are

(c1)   eq1 : sqrt(x+7)+sqrt(x+2)=1;

(d1) $$\sqrt{x+7} + \sqrt{x+2} = 1$$

(c2)   solve(eq1,x);

(d2) $$[\sqrt{x+7} = 1 - \sqrt{x+2}]$$

(c3)   eq2 : eq1 - sqrt(x+2);

(d3) $$\sqrt{x+7} = 1 - \sqrt{x+2}$$

(c4)   eq3 : expand(eq2∧2);

(d4) $$x + 7 = -2\sqrt{x+2} + x + 3$$

(c5)   eq4 : eq3 - x - 3;

(d5) $$4 = -2\sqrt{x+2}$$

(c6)   eq5 : eq4∧2;

(d6) $$16 = 4(x+2)$$

(c7)   solve(eq5,x);

(d7) $$[x = 2]$$

(c8)   subst(2,x,eq1);

(d8) $$5 = 1$$

Figure 1.10. A Macsyma 2.1 dialogue that attempts to solve Equation (1.5) by mimicking the manipulations in Equations (1.6) through (1.8).

quite involved, it may not be practical to list in detail all the capabilities of a particular command. For this reason, it is important to explore the capabilities of a CAS. Some of the exercises in this section and others throughout the book are designed with this objective in mind.

Exercises

For the exercises in this section, the following operators are useful:

- In Maple, the `diff, int, factor, solve, simplify, radsimp, subs`, and `evalf` operators (Implementation: Maple (mws)).

- In Mathematica, the `D, Integrate, Factor, Solve, Reduce, //N, Simplify, FullSimplify`, and `ReplaceAll` operators (Implementation: Mathematica (nb)).

- In MuPAD the `diff, int, Factor, solve, simplify, radsimp, subs`, and `float` operators (Implementation: MuPAD (mnb)).

1. Which of the following expressions can be factored with a CAS? Does the CAS return the result in the form you expect?

   (a) $x^2 - \dfrac{1}{4}$.

   (b) $x^2 - a^2$.

   (c) $x^2 - (\sqrt{2})^2$.

   (d) $x^2 + 1 = (x - \imath)(x + \imath)$.

   (e) $x\,y + \dfrac{1}{x\,y} + 2 = (x + 1/y)\,(y + 1/x)$.

   (f) $(\exp(x))^2 - 1 = \exp(2x) - 1$. (Notice that these two expressions are equivalent. Can a CAS factor both forms?)

   (g) $x^{2n} - 1 = (x^n - 1)(x^n + 1)$.

   (h) $x^{m+n} - x^n - x^m + 1 = (x^m - 1)(x^n - 1)$.

   (i) $x^2 + \sqrt{3}\,x + \sqrt{2}\,x + \sqrt{2}\sqrt{3}$.

   (j) $\sqrt{3}\,x^5 - \sqrt{6}\,x^4 + \sqrt{2}\,x^3 - 2x^2 + \sqrt{5}\,x - \sqrt{10} = (x - \sqrt{2}\,)(\sqrt{3}\ x^4 + \sqrt{2}\ x^2 + \sqrt{5}\,)$.

   (k) $x^4 - 10\,x^2 + 1 = (x + \sqrt{2} + \sqrt{3})(x + \sqrt{2} - \sqrt{3})(x - \sqrt{2} + \sqrt{3})(x - \sqrt{2} - \sqrt{3})$.

2. In this problem we ask you to explore the capability of a CAS to find the exact solutions to equations. Since the solution of equations is an involved operation, some computer algebra systems have either more than one command for this operation or optional parameters that modify the operation of the commands. Before attempting this exercise, you should consult the system documentation to determine best use the of the commands. In addition, a CAS may return a solution in a form that includes advanced functions that you may not be familiar with. Again, consult the system documentation for the definitions of these functions.

   Solve each of the following with a CAS.

   (a) $x^4 - 3x^3 - 7x^2 + 2x - 1 = 0$ for x. Are the roots real or do they have non-zero imaginary parts?

(b) $x^8 - 8\,x^7 + 28\,x^6 - 56\,x^5 + 70\,x^4 - 56\,x^3 + 28\,x^2 - 8\,x - 1 = 0$ for x. Since this equation has degree 8, a CAS finds the solution by using either polynomial factorization or decomposition to reduce the problem to the solution of lower degree polynomial equations. Which approach does the CAS use in this case? *Hint:* See Example 1.3.

(c) $x - \pi/2 = \cos(x + \pi)$ for x a real number. (Solution $x = \pi/2$.)

(d) $\sin(x) = 1$ for x a real. (Solution $x = \pi/2 + 2\,\pi\,n, \ \ n = 0, \pm 1, \pm 2, \ldots$.)

(e) $\sqrt{x} = 1 - x$ for $x \geq 0$. By squaring both sides of this equation we obtain the equivalent equation $x^2 - 3\,x + 1 = 0$ which has two positive roots. However, only one of these roots is a root of the original equation.

(f) $4^{(x^2)}2^x = 8$ for x a real number. By taking logarithms of both sides of this equation, we obtain the equivalent equation $2\,x^2 + x - 3 = 0$.

(g) $x^2 = 2^x$ for x a real number. (Solution $x = 2, 4, \ \ x \approx -.7666$.)

(h) $e^{x^2 - 4} + x = 3$ for x a real number. (Solution $x = 2$)

(i)   i. $\frac{x^2 - 1}{x + 1} = 2$ (Solution $x = 3$).

    ii. $x^2 - 1 = 2\,(x + 1)$ (Solution $x = 3, -1$).

Notice that (i) and (ii) are algebraically equivalent except at the point $x = -1$. (Strictly speaking, (i) is not defined at $x = -1$.) Does a CAS distinguish between these two equations?

3. Let (x, y) be the rectangular coordinates of a point in the plane, and let (r, θ) be the polar coordinates. Then

$$r^2 = x^2 + y^2, \qquad \tan(\theta) = y/x, \qquad (1.9)$$

and

$$x = r\,\cos(\theta), \qquad y = r\,\sin(\theta). \qquad (1.10)$$

(a) Can a CAS solve (1.9) for x and y?

(b) Can a CAS solve (1.10) for r and θ?

4. Use a CAS system to find the antiderivative $\int 1/\cos^5(x)\ dx$. Verify the result with a CAS by differentiation and simplification.

5. The following integral is given in an integral table

$$\int \frac{1}{(x + 1)\sqrt{x}}dx = -\arcsin\left(\frac{1 - x}{1 + x}\right), \qquad x > 0. \qquad (1.11)$$

(a) Evaluate the integral with a CAS. (All seven computer algebra systems described in Section 1.1 return a form different from Equation (1.11).)

(b) Is it possible to use a CAS to show that the antiderivative obtained in part (a) differs by at most a constant from the one given by the integral table?

6. Consider the six equations with six unknowns $\{x_1, x_2, y_1, y_2, z_1, z_2\}$:

$$
\begin{aligned}
a &= \frac{m_1 x_1 + m_2 x_2}{m_1 + m_2}, \\
b &= \frac{m_1 y_1 + m_2 y_2}{m_1 + m_2}, \\
c &= \frac{m_1 z_1 + m_2 z_2}{m_1 + m_2}, \\
r \sin(\theta) \cos(\phi) &= x_1 - x_2, \\
r \sin(\theta) \sin(\phi) &= y_1 - y_2, \\
r \cos(\theta) &= z_1 - z_2.
\end{aligned}
$$

Solve these equations with a CAS. Do you expect the solution to simplify to expressions of reasonable size?

7. Use a CAS to help find the exact value of the bounded area between the curves

$$
\begin{aligned}
u &= 2x - \frac{1}{x} + \frac{2}{x^2}, \\
v &= x + 2.
\end{aligned}
$$

Assume that $x > 0$.

8. (a) Consider the equation $x^3 - a^2 x^2 + (a+3)x - a = 0$. Use a CAS to find a real value for a so that the equation has one root of multiplicity 2 and one of multiplicity 1. *Hint:* At a root x_0 of multiplicity 2, both the polynomial and its derivative evaluate to 0.

   (b) Consider the equation $x^3 + a x^2 + a^2 x + a^3 = 0$. For $a = 0$ the equation has the root $x = 0$ with multiplicity 3. Use a CAS to show that it is impossible to find an a so that the equation has one root of multiplicity 2 and one of multiplicity 1.

9. Give a general formula for the nth derivative of the product of two functions $f(x)$ and $g(x)$. A CAS can be useful for this problem. Use a CAS to find the nth derivative of the product $f(x)g(x)$ for $n = 1, 2, 3, 4$. Use this data to find a general expression for the pattern you observe.

10. In each of the following manipulations we ostensibly show that $1 = -1$. What is the fallacy in the reasoning in each case?

    (a) $1 = \sqrt{1} = \sqrt{\imath^4} = \imath^2 = -1$ where $\imath = \sqrt{-1}$.

    (b) $1 = \sqrt{1} = \sqrt{(-1)(-1)} = \sqrt{-1}\sqrt{-1} = \imath^2 = -1$.

11. In the following manipulations we ostensibly show that every complex number is real and positive. Let $z = r e^{\imath \theta}$ be a complex number in the polar representation where $r > 0$. Certainly, if $\theta = 0$, then $z = r$ which is real and positive. If $\theta \neq 0$, then for $\alpha = e^{\imath \theta}$,

$$
\alpha^{2\pi/\theta} = \left(e^{\imath \theta}\right)^{2\pi/\theta} = e^{2\pi \imath} = \cos(2\pi) + \imath \sin(2\pi) = 1.
$$

Therefore,

$$\alpha = \left(\alpha^{2\pi/\theta}\right)^{\theta/(2\pi)} = 1^{\theta/(2\pi)} = 1.$$

Therefore $z = r$ which is real and positive. What is wrong with our reasoning?

12. Consider the following sequence of steps that ostensibly shows that $2 = 1$. Let

$$a = b. \tag{1.12}$$

Then

$$
\begin{aligned}
a^2 &= ab, \\
a^2 - b^2 &= ab - b^2, \\
(a+b)(a-b) &= b(a-b), \\
a+b &= b.
\end{aligned}
$$

Substituting Equation (1.12) into this last expression we obtain $2b = b$ and so $2 = 1$. What is the fallacy in the reasoning?

13. Consider the indefinite integral

$$\int \frac{dx}{x\ln(x)}.$$

To evaluate this integral we use the integration by parts formula $\int u\, dv = uv - \int v\, du$ with $u = 1/\ln(x)$ and $dv = dx/x$ and obtain

$$\int \frac{dx}{x\ln(x)} = 1 + \int \frac{dx}{x\ln(x)}.$$

Subtracting the integral from both side of this equation we obtain $0 = 1$. What is wrong with our reasoning?

14. Consider the system equations

$$(x^2 + y^2 + x)^2 = 9(x^2 + y^2), \tag{1.13}$$

$$x^2 + y^2 = 1. \tag{1.14}$$

(a) Solve this system of equations for x and y with a CAS.

(b) Let's try to solve this system of equations using symbol manipulation. Substituting Equation (1.14) in (1.13) we have $(x+1)^2 = 9$ and so $x = 2, -4$. Substituting $x = 2$ in (1.13) we obtain after some manipulation $y^2(y^2 + 3) = 0$ which has the real root $y = 0$. However, $x = 2, y = 0$ is not obtained by a CAS as a solution of Equation (1.14). What is the fallacy in our reasoning?

Further Reading

1.1 Computer Algebra and Computer Algebra Systems. Kline [54] gives an interesting discussion of the use of mathematics to discover new knowledge about the physical world.

Additional information on computer algebra can be found in Akritas [2], Buchberger et al. [17], Davenport, Siret, and Tournier [29], Geddes, Czapor, and Labahn [39], Lipson [64], Mignotte [66], Mignotte and Ştefănescu [67], Mishra [68], von zur Gathen and Gerhard [96], Wester [100], Winkler [101], Yap [105], and Zippel [108]. Two older (but interesting) discussions of computer algebra are found in Pavelle, Rothstein, and Fitch [77] and Yun and Stoutemyer [107].

Simon ([90] and [89]) and Wester [100] (Chapter 3) give a comparison of commercial computer algebra software. Comparisons of computer algebra systems are also found at

http://math.unm.edu/~wester/cas_review.html.

Information about computer algebra and computer algebra systems can be found at the following Internet sites.

- SymbolicNet:

http://www.SymbolicNet.org.

- Computer Algebra Information Network (CAIN):

http://www.riaca.win.tue.nl/CAN/.

- COMPUTER ALGEBRA, Algorithms, Systems and Applications:

http://www-troja.fjfi.cvut.cz/~liska/ca/.

- sci.math.symbolic discussion site:

http://mathforum.org/discussions/about/sci.math.symbolic.html

The Association for Computing Machinery (ACM) has a *Special Interest Group on Symbolic and Algebraic Manipulation (SIGSAM)*. This group publishes a quarterly journal the *SIGSAM Bulletin* which provides a forum for exchanging ideas about computer algebra. In addition, SIGSAM sponsors an annual conference, the *International Symposium on Symbolic and Algebraic Computation (ISSAC)*. Information about SIGSAM can be found at the Internet site

http://www.acm.org/sigsam.

The main research journal in computer algebra is the *Journal of Symbolic Computation* published by Academic Press. Information about this journal can be found at

http://www.academicpress.com/jsc.

Computers have also been used to prove theorems. See Chou [20] for an introduction to computer theorem proving in Euclidean geometry.

Computers have even been used to generate mathematical conjectures or statements which have a high probability of being true. See Cipra [21] for details.

There has also been some work to use artificial intelligence symbolic programs to help interpret the results of numerical computer experiments and even to suggest which experiments should be done. See Kowalik [58] for the details.

See Kajler [51] for a discussion of research issues in human-computer interaction in symbolic computation.

1.2 Applications of Computer Algebra. The article by Nowlan [76] has a discussion of the consequences a purely mechanical approach to mathematics. Stoutemyer [94], which describes some problems that arise with CAS software, should be required reading for any user of this software.

Bernardin (see [7] or [8]) compares the capability to solve equations for six computer algebra systems. Some of the equations in Exercise 2 on page 22 are from these references.

Exercise 11 on page 24 is from *The College Mathematics Journal*, Vol. 27, No. 4, Sept. 1996, p. 283. This journal occasionally has examples of faulty symbolic manipulation in its section *Fallacies, Flaws, and Flimflam.* See

http://www.maa.org/pubs/cmj.html.

2

Elementary Concepts of Computer Algebra

In this chapter we introduce a language that is used throughout the book to describe the concepts, examples, and algorithms of computer algebra. The language is called *mathematical pseudo-language* or simply MPL. In Sections 2.1 and 2.2 we describe the form of an MPL mathematical expression and discuss what happens to an expression during the evaluation process. In Section 2.3 we consider elementary MPL programs and give a case study that illustrates the concept. Finally, in Section 2.4 we describe MPL lists and sets, which are two ways to represent collections of mathematical expressions.

2.1 Mathematical Pseudo-language (MPL)

Mathematical pseudo-language (MPL) is a symbolic language that is used in this book to describe the concepts, examples, and algorithms of computer algebra. The term *pseudo-language* is used to emphasize that MPL is not a real CAS language that has been implemented on a computer. Although MPL is similar in spirit to real computer algebra languages, it is less formal and utilizes both mathematical symbolism and ordinary English when appropriate. The reader should have little difficulty following discussions in MPL.

The reader may wonder, why introduce another algorithmic language? Why not use the programming language associated with a particular CAS?

One reason has to do with the current state of language and system development in the computer algebra field. There is now a proliferation of computer algebra systems, and, undoubtedly, there will be new ones in the future. Each system has its strong points and limitations, and its own following among members of the technical community. The systems are distinguished from each other by the nature of the mathematical knowledge encoded in the system and the language facilities that are available to access and extend this knowledge. However, at the basic level, there are more similarities than differences, and the organization of mathematical concepts and language structures do not differ significantly from system to system. By using a generic pseudo-language we are able to emphasize the concepts and algorithms of symbolic computation without being confined by the details, quirks, and limitations of a particular language.

Perhaps the most important role for MPL is that it provides a way to evaluate and compare computer algebra systems and languages. In fact, a useful approach to this chapter is to read it with one or more computer algebra systems at your side and, as MPL concepts and operations are described, implement them in real software. Although you will find that MPL's style is similar to real software, you will also find differences between it and real languages, and especially subtle differences between the languages themselves.

Mathematical Expressions in MPL

To use a computer algebra system effectively, it is important to have a clear understanding of both the structure and meaning of mathematical expressions. Since there is much to say about this subject, mathematical expressions will occupy much of our attention in this chapter and Chapter 3. In Chapter 4 we introduce other elements of the MPL language.

Let's begin by looking at the various forms an MPL expression can have. Roughly speaking, MPL expressions are similar to those found in ordinary mathematical symbolism with some allowance made to accommodate the need for more precision in a computational environment. MPL expressions are constructed using the following symbols and operators:

Integers and fractions. Software that performs the exact manipulation of mathematical expressions must have the capability to perform exact arithmetic. Real floating point arithmetic, which is used by conventional programming languages for purely numerical work, involves round-off error and is not appropriate for most computer algebra computation. Indeed, even the small numerical errors that are inevitable with floating point arithmetic can alter the mathematical properties of an expression. To illustrate this point, consider the following two expressions which are identical except for a small change in one coefficient:

$$f = \frac{x^2 - 1}{x - 1}, \qquad g = \frac{x^2 - .99}{x - 1}.$$

Although the numerical values of f and g are nearly the same for most values of x, the mathematical properties of the two expressions are different. First of all, f simplifies to the polynomial $x + 1$ when $x \neq 1$ while g does not. Consequently, their antiderivatives differ by a logarithmic term:

$$\int f \, dx = x^2/2 + x + C, \qquad x \neq 1,$$

$$\int g \, dx = x^2/2 + x + .01 \ln(x - 1) + C, \qquad x \neq 1.$$

Furthermore, the graph of g has an asymptote at $x = 1$, while f is simply undefined at $x = 1$.

To avoid these discrepancies, MPL utilizes *exact arbitrary precision rational number arithmetic* for most numerical computations rather than approximate floating point arithmetic. The term *arbitrary precision* means an integer or fraction can have an arbitrary number of digits. Examples include

$$2/3, \quad -1/4, \quad 123456789/987654321, \quad 24329020008176640000.$$

Arithmetic calculations are performed using the ordinary rules for rational number arithmetic.

All computer algebra systems utilize this type of arithmetic, however, because a computer is a finite machine, there is a maximum number of digits permitted in a number. This bound is usually quite large and rarely a limitation in applications.

Real numbers. In MPL, a real number is one that has a finite number of digits, includes a decimal point, and may include an optional power of 10. Examples include

$$467.22, \quad .33333333, \quad 6.02 \cdot 10^{23}. \tag{2.1}$$

Real number arithmetic is similar to real floating point arithmetic in a conventional programming language. Since this mode of computation may involve round-off error, it is, in general, inexact. Most computer algebra systems support real numbers, and some systems allow for choice of numerical precision.

The definition of an MPL real number should not be confused with the mathematical concept of a real number. Since all MPL real numbers have a finite number of digits, they are really rational numbers in the mathematical sense. In mathematics, a real number that is not rational is called an *irrational number*. For example, $\sqrt{2}$, π, and e are irrational numbers, and it is hard to imagine doing symbolic computation without them. Since irrational numbers require an infinite decimal representation which is not possible in a computational setting, they are represented instead using reserved symbols (e, π), algebraic expressions $(2 \wedge (1/2))$, or function forms $(\ln(2))$, all of which are described below.

Identifiers. In MPL, an *identifier* is a string of characters constructed with English letters, Greek letters, the digits $0, 1, \ldots, 9$, and the underscore symbol "_". An identifier begins with an English or Greek letter. The following are examples of MPL identifiers:

$$x, \quad y1, \quad \alpha, \quad general\_solution, \quad \Delta x.$$

Identifiers are used in MPL as *programming variables* that represent the result of a computation, as function, operator, or procedure names, as *mathematical symbols* that represent indeterminates (or variables) in a mathematical expression, and as reserved symbols. All computer algebra languages use identifiers in this way although the characters allowed in an identifier name vary from system to system.

Algebraic operators and parentheses. The algebraic operators in MPL are listed in Figure 2.1. Parentheses are used as they are in mathematics to alter the structure of an expression. Examples of expressions that include the operators, numbers, and identifiers described so far are

$$(n - m)!, \quad x \wedge 2 - 5 * x + 6, \quad ((x + \Delta x) \wedge 2 - x \wedge 2)/\Delta x.$$

Mathematical Operation	MPL Operator
addition, subtraction	$+, \; -$
multiplication, division	$*, \; /$
power	$\wedge$
factorial	!

Figure 2.1. Algebraic operators in MPL.

Reserved symbols. A *reserved symbol* is an identifier or other mathematical symbol that has mathematical meaning. In MPL, the reserved symbols include π, e, $\imath$ (for $\sqrt{-1}$), ∞, and the logical constants **true** and **false**.

MPL	Maple	Mathematica	MuPAD
π	Pi	Pi	PI
e	exp(1)	E	E
$\imath$	I	I	I
∞	infinity	Infinity	infinity
true	true	True	TRUE
false	false	False	FALSE

Figure 2.2. MPL reserved symbols in Maple, Mathematica, and MuPAD.

A few more reserved symbols are introduced in later sections. The corresponding reserved symbols in three computer algebra systems are given in Figure 2.2.

In a CAS, reserved symbols acquire mathematical meaning through the actions of the transformation rules encoded in the system. For example, most computer algebra systems recognize the simplifications

$$\sin(\pi/2) \quad \to \quad 1, \tag{2.2}$$

$$\arctan(1) \quad \to \quad \pi/4, \tag{2.3}$$

$$\ln(e \wedge 2) \quad \to \quad 2, \tag{2.4}$$

$$\imath \wedge 2 \quad \to \quad -1,$$

$$e \wedge (-\imath * \pi) \quad \to \quad -1$$

as either part of the evaluation process or the output of a simplification operator. (Implementation: Maple (mws), Mathematica (nb), MuPAD (mnb).)

Function forms. In MPL, function forms are used for mathematical functions ($\sin(x)$, $\exp(x)$, $\arctan(x)$, etc.), mathematical operators (*Expand*(u), *Factor*(u), *Integral*(u, x), etc.), and undefined functions ($f(x)$, $g(x, y)$, etc.).

In a CAS, mathematical functions acquire meaning through the actions of transformation rules encoded in the system. For example, most computer algebra systems obtain function transformations similar to the simplifications (2.2), (2.3), and (2.4) above.

Function forms that manipulate and analyze mathematical expressions are called *mathematical operators*. Although computer algebra systems contain hundreds of mathematical operators, we use only a small number of them in this book. Figure 2.3 gives some of the MPL operators used in the examples, algorithms, and exercises in this chapter, and Figure 2.4

Mathematical Operation	MPL Operator	Example
Absolute value, $|u|$	$Absolute\_value(u)$	$Absolute\_value(-2) \to 2$
Evaluate rational numbers, arithmetic operations, and numerical functions in an expression u to a real value	$Decimal(u)$	$Decimal(1/4) \to .25$ $Decimal(x + 1/4) \to x + .25$ $Decimal(\sin(2) + 1/2) \to 1.409297$
Substitution in u of each occurrence of t by r	$Substitute(u, t = r)$	$Substitute(2 * x + 1, x = b + 1)$ $\to 2(b + 1) + 1$
ith operand in an expression u	$Operand(u, i)$	$Operand(a + b + c, 2) \to b$ $Operand(\{a, b, c\}, 3) \to c$ $Operand(a = b, 2) \to b$
Degree in x of a polynomial expression u	$Degree(u, x)$	$Degree(x \wedge 2 + 5 * x + 7, \ x)$ $\to 2$
Coefficient of x^j in a polynomial expression u	$Coefficient(u, x, j)$	$Coefficient(x \wedge 2 + 5 * x + 7, \ x, \ 1)$ $\to 5$
Algebraic expansion	$Algebraic\_expand(u)$	$Algebraic\_expand($ $(x + 2) * (x + 3))$ $\to x^2 + 5 * x + 6$
Polynomial factorization	$Factor(u)$	$Factor(x \wedge 2 + 5 * x + 6)$ $\to (x + 2)(x + 3)$
Solution of an equation u for x	$Solve(u, x)$	$Solve(a * x = b, x)$ $\to x = b/a$
or a set of equations for a set of variables	$Solve(\{u_1, \ldots, u_n\}, \{x_1, \ldots, x_n\})$	$Solve($ $\{2 * x + 4 * y = 3, \ 3 * x - y = 7\},$ $\{x, y\})$ $\to \{x = 31/14, \ y = -5/14\}$
$\lim\limits_{x \to a} u$	$Limit(u, x, a)$	$Limit(1/x, x, \infty) \to 0$
$\dfrac{du}{dx}$	$Derivative(u, x)$	$Derivative(\sin(x), x) \to \cos(x)$
$\int u \, dx$	$Integral(u, x)$	$Integral(\cos(x), x) \to \sin(x)$
Solution of a differential equation u for $y(x)$	$Solve\_ode(u, x, y)$	$Solve\_ode($ $Derivative(y(x), x) = y(x), x, y)$ $\to C * \exp(x)$

Figure 2.3. Some mathematical operators in MPL. In column 3, the expression to the right of the evaluation symbol $\to$ is the result obtained by evaluating the operator. The corresponding operators in three computer algebra systems are giving in Figure 2.4.

MPL	Maple	Mathematica	MuPAD
$Absolute\_value(u)$	abs(u)	Abs[u]	abs(u)
$Decimal(u)$	evalf(u)	u//N	float(u)
$Substitute(u, t = r)$	subs(t=r,u)	ReplaceAll[u,t->r]	subs(u,t=r)
$Operand(u, i)$	op(i,u)	Part[u,i]	op(u,i)
$Degree(u, x)$	degree(u,x)	Exponent[u,x]	degree(u,x)
$Coefficient(u, x, j)$	coeff(u,x,j)	Coefficient[u,x,j]	coeff(u,x,j)
$Algebraic\_expand(u)$	expand(u)	Expand[u]	expand(u)
$Factor(u)$	factor(u)	Factor[u]	expr(factor(u))
$Solve(u, x)$	solve(u,x)	Solve[u,x]	solve(u,x)
$Limit(u, x, a)$	limit(u,x=a)	Limit[u,x->a]	limit(u,x,a)
$Derivative(u, x)$	diff(u,x)	D[u,x]	diff(u,x)
$Integral(u, x)$	int(u,x)	Integrate[u,x]	int(u,x)
$Solve\_ode(u, x, y)$	dsolve(u,y(x))	DSolve[u,y[x],x]	solve(ode(u,y(x)))

Figure 2.4. The operators in the Maple, MuPAD, and Mathematica systems that correspond most closely to the MPL operators in Figure 2.3. (Implementation: Maple (mws), Mathematica (nb), MuPAD (mnb).)

gives the operators in the Maple, Mathematica, and MuPAD systems that correspond most closely to these operators.

Another important function form is the *undefined function* which is an expression in function notation (e.g., $f(x)$, $g(x, y)$, $h(n + 1)$), where the function is undefined. In a computational setting this means there are no transformation rules or other properties associated with the function beyond the implied dependence of the function name on the expressions in parentheses. In ordinary mathematical notation, dependency relationships of this sort are usually understood from context. In the computational setting, however, more precision is required, and undefined functions provide one way to represent this dependency.

One use of undefined functions is in expressions that involve arbitrary or unknown functions. For example, in the differentiation

$$Derivative(f(x) * g(x),\ x) \rightarrow \frac{df(x)}{dx}\,g(x) + f(x)\,\frac{dg(x)}{dx},$$

MPL's *Derivative* operator uses the dependency information to obtain a general form of the product rule. Without this information, the *Derivative* operator assumes that f and g do not depend on x, and so *Derivative*($f *$ g, x) evaluates to 0.

All computer algebra systems use function forms in the three ways described above. In Figure 2.5, we give a Mathematica dialogue which obtains the solution of the differential equation

$$\frac{dy}{dx} + y = x + \exp(-2x) \tag{2.5}$$

$In[1] := u = D[y[x], x] + y[x] == x + \text{Exp}[-2*x]$

$Out[1] = y[x] + y'[x] == e^{-2x} + x$

$In[2] := \text{DSolve}[u, y[x], x]$

$Out[2] = \{\{y[x] \to e^{-x}(-e^{-x} + e^{x}(-1 + x)) + e^{-x} C[1]\}\}$

Figure 2.5. A Mathematica dialogue which obtains the solution of a differential equation. The Mathematica language uses the brackets [and] to represent function forms. (Implementation: Maple (mws), Mathematica (nb), MuPAD (mnb).)

which illustrates this point. At the first prompt *In[1]*, we enter the differential equation using the function notation y[x] to represent the dependency of y on x, and at *Out[1]*, the system returns an expression where the derivative is represented in symbolic form as y'[x]. At *In[2]*, we enter the command to solve the differential equation and obtain the general solution in *Out[2]*. Observe that Mathematica represents the arbitrary constant in the solution by C[1].

Relational operators and expressions. In MPL, a relational expression is one that expresses a relationship between two expressions using one of the relational operators

$$=, \ \neq, \ <, \ \leq, \ >, \ \geq \ .$$

Examples include $x \wedge 2 + 2*x - 1 = 0$, $i < n$, and $\Delta p * \Delta x \geq h$.

Logical operators and expressions. An MPL logical expression is one constructed using logical constants (**true** and **false**), relational expressions, and identifiers combined together with one or more of the logical operators **and, or,** and **not.** As with algebraic expressions, parentheses are used to alter the structure of an expression. Examples include

(**true and false**) **or true,**   **not** (p **and** q), $0 \leq x$ **and** $x \leq 1$.

All computer algebra languages provide relational and logical expressions (see Figure 2.6) although their roles in the languages vary from system to system[1]

---

[1] We return to this point in Section 3.2 (see pages 97-99).

MPL	Maple	Mathematica	MuPAD
true	true	True	TRUE
false	false	False	FALSE
$=$	=	== or ===	=
$\neq$	<>	!= or =!=	<>
$<$	<	<	<
$\leq$	<=	<=	<=
$>$	>	>	>
$\geq$	>=	>=	>=
and	and	&&	and
or	or	||	or
not	not	!	not

Figure 2.6. Relational operators, logical constants, and logical operators in Maple, Mathematica, and MuPAD.

Sets and lists. In MPL, both sets and lists are used to represent collections of mathematical expressions. A set is expressed using the braces { and } and a list using the brackets [and]. Examples include

$$\{2*x+4*y=3,\ 3*x-y=7\}, \quad [1,\ x,\ x\wedge 2\ , x\wedge 3].$$

In MPL, a set or a list is considered a mathematical expression rather than a data structure that contains mathematical expressions[2]. In fact, a set or a list can be a sub-expression of another mathematical expression. For example, the expression

$$Solve(\{2*x+4*y=3,\ 3*x-y=7\},\ \{x,y\})$$

which contains sets, is used to obtain the solution of a system of linear equations. Although both sets and lists are used for collections of expressions, they have different mathematical properties and are used in different ways in our examples and algorithms. In Section 2.4, we discuss these differences and describe the operations that are appropriate for each of them.

Most computer algebra languages provide lists and sets.

MPL mathematical expressions. An MPL mathematical expression is any valid mathematical expression that is formed using integers, fractions, real numbers, identifiers, reserved symbols, function forms, sets, lists, and the algebraic, relational and logical operators described above. (A few additional operators are introduced in later sections.) For our purposes, any expression with appropriate operands for each operator and

---

[2] A data structure is a programming language structure that is used to organize data. An array is an example of a data structure.

balanced parentheses is valid. For example, $\{a, b, c\} * y$ is not valid because the first operand of $*$ is not appropriate. The expression $(a + b) * c)$ is not valid because there is a dangling right parenthesis.

Although this description of mathematical expressions is sufficient for our purposes, a theoretician would rightfully complain that we haven't given a definition at all since the word *valid* is not precisely defined. A more formal definition would include a set of *syntax* or *grammar* rules that define when a sequence of symbols is a valid expression in our language. The syntax rules would tell us, for example, that the expression $m * x + b$ is a valid expression, while $(a + b) * c)$ is not.

The syntax rules for expressions are quite involved, even for expressions as simple as those considered here. A precise listing of the rules is essential for the designer of a computer algebra system, who must determine which expressions are valid statements in a language. The syntax rules are encoded in a program called a *parser* that determines if an input expression is a valid expression in the language, determines its structure, and translates it into an *internal form* that is used by the CAS to manipulate and analyze the expression. The *structure* of an expression involves the relationships between the operators and operands that make up the expression. For example, the expression $m * x + b$ has the structure of a sum with operands $m * x$ and b rather than a product with operands m and $x + b$.

Although syntax rules and parsing algorithms are important topics for system design, they are not essential to the understanding of computer algebra and are not addressed in this book. On the other hand, since an understanding of expression structure is essential for computer algebra programming, we examine this topic in detail in Chapter 3.

Variable Initialization and Assignment

In MPL (as in a CAS), all variables are initially undefined symbols. This assumption allows a variable to fulfill its traditional role as an indeterminate symbol in a mathematical expression.

A variable that is used in the programming sense to represent the result of a computation is given a value with an *assignment* statement. In MPL, the assignment operator is a colon followed by an equal sign (:=), and an assignment statement has the form:

$$\text{variable} := \text{mathematical expression}.$$

An assignment statement causes two actions to occur. First, the expression to the right of the assignment symbol is evaluated giving a new expression.

Next, this new expression is assigned to the variable to the left of the assignment symbol. For example in

$$y := Factor(x \wedge 2 + 5 * x + 6)$$

the right side evaluates to $(x+2)*(x+3)$, which is assigned to the variable y. In future manipulations, this expression is the value of y.

All computer algebra languages provide assignment statements that operate in this way[3].

Role of Mathematical Expressions in MPL

One aspect of computer algebra programming that distinguishes it from conventional programming is the role of mathematical expressions. In MPL, mathematical expressions have two (somewhat overlapping) roles as either *program statements* that represent a computational step in a program or as *data objects* that are processed by program statements. For example, suppose x is an unassigned variable and consider the statement

$$f := x \wedge 2 + 5 * x + 6. \tag{2.6}$$

In this statement the polynomial expression which is assigned to f is a data object that can be manipulated or analyzed by other program statements. On the other hand, in

$$\begin{aligned} g \quad := \quad & Substitute(Derivative(f, x), x = c) \; * \; (x - c) \\ + \; & Substitute(f, x = c), \end{aligned} \tag{2.7}$$

the expression to the right of the assignment operator is a program statement that obtains the formula for the tangent line to f at $x = c$, and assigns the result to the variable g. For example, if c is assigned the expression $1/2$ and f is given by the statement (2.6), then g is assigned the new data object, the expression $6 * (x - 1/2) + 35/4$.

Although this description of the role of expressions is useful for emphasizing their dual nature, the distinction should not be taken too literally. Indeed, the role of an expression can depend on other actions that have occurred in a computation. For example, in Statement (2.6), the polynomial is a data object as long as x has not been assigned. On the other hand, if x has been assigned the integer 3, the polynomial in Statement (2.6) can be viewed as a program statement which upon evaluation obtains the expression 30 which is then assigned to f.

---

[3]In Maple and MuPAD, the assignment symbol is the colon followed by the equal sign (:=); in Mathematica, the assignment symbol is the equal sign (=).

$<1>Algebraic\_expand((x+2)^2 * (x+3));$

$$\rightarrow \quad x^3 + 7x^2 + 16x + 12$$

$<2> Factor(2 * x^3 + 7 * x^2 * y + 4 * x^2 + 14 * x * y + 18 * x + 63 * y);$

$$\rightarrow \quad (x^2 + 2x + 9)(2x + 7y)$$

$<3> Integral(x * \sin(x),\ x);$

$$\rightarrow \quad \sin(x) - x\cos(x)$$

Figure 2.7. An MPL dialogue. (Implementation: Maple (mws), Mathematica (nb), MuPAD (mnb).)

Most computer algebra languages employ expressions as both program statements and data objects, although a language may restrict the use of some expression types to certain contexts[4].

MPL Dialogues

An MPL dialogue which mimics the interactive dialogues found in real computer algebra systems is given in Figure 2.7. In this simulation, the prompt is represented by a positive integer surrounded by the symbols $<$ and $>$ and the mathematical expression following each prompt represents an input to our imaginary system. Following the practice in some computer algebra systems, each input statement is terminated by a semicolon[5]. The arrow "$\rightarrow$" to the left of the centered expressions means "evaluates to" and indicates the result of evaluating the preceding input expression.

MPL Notation versus Ordinary Mathematical Notation

The notation for MPL expressions closely resembles the notation used for input expressions in most computer algebra systems. However, as with most programming notations, it is notoriously unreadable for large expressions. On the other hand, ordinary mathematical notation, which is

---

[4]The Maple, Mathematica and MuPAD systems allow all the expressions described here both as program statements and data objects. On the other hand, the Macsyma system does not permit some logical expressions as data objects. For example, the logical expression **p and q**, with **p** and **q** undefined symbols, cannot be entered in the interactive mode in that system.

[5]The Mathematica system does not require a termination symbol at the end of an expression. Most systems allow a choice of terminating symbol to provide an option to display or not display a result.

far more understandable, lacks the precision of MPL notation and is un-suitable in some computational contexts. Since there is clearly a place for both notations, we adopt the following strategy for using and intermingling the two:

- We usually use MPL notation for input to MPL dialogues and for statements, procedures, and examples that involve manipulations in a computational context. However, in some of these situations, MPL notation is unwieldy and for clarity we resort to ordinary mathematical notation. For example, in the MPL dialogue in Figure 2.7, we use raised exponents for powers in the inputs <2> and <3> instead of using the $\wedge$ operator.

- We usually use ordinary mathematical notation in theorems, examples, and discussions that are not in a computational context. In addition, since most computer algebra systems display output in a form similar to ordinary mathematical notation, we use this form for output in MPL dialogues as well (see Figure 2.7). There are, however, some instances where the conciseness of MPL notation invites its use in purely mathematical contexts.

We assume the reader can readily translate between the two notations.

Translating Mathematical Discourse into MPL

We conclude this section with an example that shows how a sequence of operations in ordinary mathematical discourse is translated into a sequence of statements in MPL.

Example 2.1. Consider the following equation which defines y implicitly as a function of x:

$$\exp(x) + y^4 = 4\,x^2 + y. \tag{2.8}$$

Let's consider the manipulations that are used to compute implicitly the derivatives

$$\frac{dy}{dx} \text{ and } \frac{d^2y}{dx^2}.$$

First, differentiating both sides of Equation (2.8) with respect to x, we have

$$\exp(x) + 4\,y^3\frac{dy}{dx} = 8\,x + \frac{dy}{dx}. \tag{2.9}$$

Solving for $\frac{dy}{dx}$, we obtain

$$\frac{dy}{dx} = \frac{-\exp(x) + 8\,x}{4\,y^3 - 1}. \tag{2.10}$$

To obtain the second derivative, we differentiate this expression

$$\frac{d^2 y}{dx^2} = \frac{-\exp(x) + 8}{4\,y^3 - 1} - 12\,\frac{\left(-\exp(x) + 8\,x\right) y^2 \frac{dy}{dx}}{\left(4\,y^3 - 1\right)^2}, \tag{2.11}$$

and then substitute the right side of Equation (2.10) for $\frac{dy}{dx}$ to obtain

$$\frac{d^2 y}{dx^2} = \frac{-\exp(x) + 8}{4\,y^3 - 1} - 12\,\frac{\left(-\exp(x) + 8\,x\right)^2 y^2}{\left(4\,y^3 - 1\right)^3}. \tag{2.12}$$

Let's consider now the MPL operations that produce the manipulations in Equations (2.8) through (2.12). Three operations are required: differentiations in (2.9) and (2.11), a solution of a linear equation in (2.9), and a substitution in (2.12). These manipulations are readily translated into a sequence of statements in MPL (see Figure 2.8). We begin at <1> by assigning Equation (2.8) to u, where an undefined function $y(x)$ is used to represent the dependence of y on x. At statement <2>, we use the *Derivative* operator to differentiate both sides of u and assign this result

---

<1> $u := \exp(x) + y(x)^4 = 4\,x^2 + y(x);$

$\rightarrow \qquad \exp(x) + y(x)^4 = 4\,x^2 + y(x)$

<2> $v := Derivative(u,\, x);$

$\rightarrow \qquad \exp(x) + 4\,y(x)^3\,\frac{dy(x)}{dx} = 8\,x + \frac{dy(x)}{dx}$

<3> $First\_derivative := Solve(v,\, Derivative(y(x), x));$

$\rightarrow \qquad \frac{dy(x)}{dx} = \frac{-\exp(x) + 8\,x}{4\,y(x)^3 - 1}$

<4> $w := Derivative(First\_derivative,\, x);$

$\rightarrow \qquad \frac{d^2 y(x)}{dx^2} = \frac{-\exp(x) + 8}{4\,y(x)^3 - 1} - 12\,\frac{\left(-\exp(x) + 8\,x\right) y(x)^2 \frac{dy(x)}{dx}}{\left(4\,y(x)^3 - 1\right)^2}$

<5> $Second\_derivative := Substitute(w,\, First\_derivative);$

$\rightarrow \qquad \frac{d^2 y(x)}{dx^2} = \frac{-\exp(x) + 8}{4\,y(x)^3 - 1} - 12\,\frac{\left(-\exp(x) + 8\,x\right)^2 y(x)^2}{\left(4\,y(x)^3 - 1\right)^3}$

Figure 2.8. The MPL manipulations that correspond to Equations (2.8) through (2.12).

to v. At <3>, we use the *Solve* operator to solve the equation v for the expression

$$\frac{dy(x)}{dx}$$

and assign this result to *First_derivative*. To obtain the second derivative, at <4> we apply the *Derivative* operator to both sides of *First_derivative* and assign this result to w. Finally, at <5> we use the *Substitute* operator to apply the substitution defined by *First_derivative* to w and assign this result to *Second_derivative*. □

Interactive Dialogues with Real Computer Algebra Systems

MPL provides a way to express a sequence of symbolic calculations in a form that resembles the statements and operations in a real CAS. Although the MPL dialogues indicate in a general way the sequence of manipulations needed for a calculation, we caution the reader not to take the input statements and outputs in the dialogues too literally. The actual implementation of a dialogue in a real CAS language will vary from system to system.

To illustrate this point, we implement the MPL dialogue in Figure 2.8 in the Maple, Mathematica, MuPAD languages (see Figures 2.9, 2.10, and 2.11). These dialogues use each system's versions of MPL's *Derivative*, *Solve*, and *Substitute* operators along with each system's version of MPL's selection operator *Operand*(u, i) which returns the ith operand of the expression u. For example, this operator obtains

$$Operand(a + b + c, 2) \rightarrow b,$$
$$Operand(\{a, b, c\}, 3) \rightarrow c,$$
$$Operand(a = b, 2) \rightarrow b.$$

Although this operator was not needed in the MPL dialogue, it is required to handle the various forms of the input and output of the *Solve* and *Substitute* operators in a real CAS.

Maple

The Maple implementation of Figure 2.8 is given in Figure 2.9. The statements at the first two prompts are similar to those in the MPL dialogue, although Maple displays the results as assignments and displays the derivative with partial derivative notation. At the third prompt, Maple solves

```
> u := exp(x)+y(x)^4=4*x^2+y(x);
```

$$u := e^x + y(x)^4 = 4\,x^2 + y(x)$$

```
> v := diff(u,x);
```

$$v := e^x + 4\,y(x)^3\left(\frac{\partial}{\partial x}\,y(x)\right) = 8\,x + \left(\frac{\partial}{\partial x}\,y(x)\right)$$

```
> d := solve(v,diff(y(x),x));
```

$$d := \frac{-e^x + 8\,x}{4\,y(x)^3 - 1}$$

```
> First_derivative := diff(y(x),x) = d;
```

$$\mathit{First\_derivative} := \frac{\partial}{\partial x}\,y(x) = \frac{-e^x + 8\,x}{4\,y(x)^3 - 1}$$

```
> w := diff(First_derivative,x);
```

$$w := \frac{\partial^2}{\partial x^2}\,y(x) = \frac{-e^x + 8}{4\,y(x)^3 - 1} - \frac{12\,(-e^x + 8\,x)\,y(x)^2\,(\frac{\partial}{\partial x}\,y(x))}{(4\,y(x)^3 - 1)^2}$$

```
> subs(First_derivative,w);
```

$$\frac{\partial}{\partial x}\frac{-e^x + 8\,x}{4\,y(x)^3 - 1} = \frac{-e^x + 8}{4\,y(x)^3 - 1} - \frac{12\,(-e^x + 8\,x)^2\,y(x)^2}{(4\,y(x)^3 - 1)^3}$$

```
> Second_derivative := diff(y(x),x,x) = subs(First_derivative,op(2,w));
```

$$\mathit{Second\_derivative} := \frac{\partial^2}{\partial x^2}\,y(x) = \frac{-e^x + 8}{4\,y(x)^3 - 1} - \frac{12\,(-e^x + 8\,x)^2\,y(x)^2}{(4\,y(x)^3 - 1)^3}$$

Figure 2.9. A Maple implementation of the MPL dialogue in Figure 2.8. (Implementation: Maple (mws).)

the equation v for `diff(y(x),x)`, where the solution is returned as an expression

$$\frac{-e^x + 8\,x}{4\,y(x)^3 - 1}, \tag{2.13}$$

rather than, as in the MPL dialogue, as an equation with the derivative symbol on the left side. We compensate for this at the fourth prompt by entering an equation with the derivative symbol on the left side. At the fifth prompt, we differentiate both sides of the equation `First_derivative`, and at the sixth prompt apply Maple's `subs` command to substitute Expression (2.13) for the first derivative symbol in the previous expression. Unfortunately, we get a little more than we bargained for, since the left side of the equation is returned as a first derivative symbol applied to an expression rather than as a second derivative symbol. The reason for this has to do with Maple's internal representation of the second derivative symbol as nested first derivatives

$$\frac{\partial}{\partial x}\left(\frac{\partial}{\partial x}\,y(x)\right).$$

Since Maple's `subs` operator replaces all occurrences of the first derivative symbol with Expression (2.13), we obtain the result shown in the dialogue. Finally, at the seventh prompt, we compensate for this by using Maple's operator `op` (which selects operands of an expression) to select the right side of `w` and by applying the `subs` operator to the resulting expression. In addition, to obtain the MPL result, we include the second derivative symbol on the left side of an equation.

Mathematica

The Mathematica implementation of Figure 2.8 is given in Figure 2.10. The statements at $In[1]$ and $In[2]$ are similar to those in the **MPL** dialogue. Observe that Mathematica uses the equal sign (=) for assignment, two equal signs (==) for an equal sign in an equation, and the D operator for differentiation. At $In[3]$, Mathematica's `Solve` operator is used to solve the equation v for the derivative, where the result is returned as a set which contains another set which contains the solution. The expression

$$y'[x] \;\rightarrow\; \frac{-e^x + 8x}{-1 + 4\,y[x]^3}, \tag{2.14}$$

which is known as a *transformation rule* in the Mathematica language, is the form Mathematica uses for the substitution operation later in the dialogue. However, if we insist that the solution be displayed as an equation, we can obtain this form by using Mathematica's `Part` operator which selects operands of an expression. At $In[4]$, the expression `Part[s,1]` removes the outer set braces, the next **Part** operation removes the inner set braces, and the outer **Part** operation selects the right side of Expression (2.14). We obtain the desired form by entering an equation with the derivative symbol on the left side and then assign the result to `FirstDerivative`[6].

---

[6]Since the underscore character (_) has special meaning in Mathematica, we use the identifiers `FirstDerivative` and `SecondDerivative` instead of the identifiers *First_derivative* and *Second_derivative* used in the MPL dialogue.

$In[1] := u = exp[x] + y[x]^4 == 4 * x^2 + y[x]$

$Out[1] = e^x + y[x]^4 == 4x^2 + y[x]$

$In[2] := v = D[u, x]$

$Out[2] = e^x + 4 y[x]^3 y'[x] == 8x + y'[x]$

$In[3] := s = Solve[v, D[y[x], x]]$

$Out[3] = \{\{y'[x] \to \dfrac{-e^x + 8x}{-1 + 4 y[x]^3}\}\}$

$In[4] := FirstDerivative = D[y[x], x]] == Part[Part[Part[s, 1], 1], 2]$

$Out[4] = y'[x] == \dfrac{-e^x + 8x}{-1 + 4 y[x]^3}$

$In[5] := w = D[FirstDerivative, x]]$

$Out[5] = y''[x] == \dfrac{8 - e^x}{-1 + 4 y[x]^3} - \dfrac{12 (-e^x + 8x) y[x]^2 y'[x]}{(-1 + 4 y[x]^3)^2}$

$In[6] := SecondDerivative = ReplaceAll[w, Part[Part[s, 1], 1]]$

$Out[6] = y''[x] == -\dfrac{12 (-e^x + 8x)^2 y[x]^2}{(-1 + 4 y[x]^3)^3} + \dfrac{8 - e^x}{-1 + 4 y[x]^3}$

Figure 2.10. A Mathematica implementation of the MPL dialogue in Figure 2.8. (Implementation: Mathematica (nb).)

To obtain the second derivative, at $In[5]$ we differentiate both sides of the equation FirstDerivative, and at $In[6]$, we obtain the substitution with Mathematica's ReplaceAll command. The substitution is defined by Part[Part[s,1],1] which selects the expression (2.14).

MuPAD

The MuPAD implementation of Figure 2.8 is given in Figure 2.11. The statements at the first two prompts are similar to those in the MPL dia-

- $u := \exp(x) + y(x) \wedge 4 = 4 * x \wedge 2 + y(x);$

  $$e^x + y(x)^4 = y(x) + 4 \cdot x^2$$

- $v := \text{diff}(u, x);$

  $$e^x + 4 \cdot y(x)^3 \cdot \frac{\partial}{\partial x} y(x) = 8 \cdot x + \frac{\partial}{\partial x} y(x)$$

- $v2 := \text{subs}(v, \text{diff}(y(x), x) = Dy);$

  $$e^x + 4 \cdot Dy \cdot y(x)^3 = 8 \cdot x + Dy$$

- $d := \text{solve}(v2, Dy, \text{IgnoreSpecialCases});$

  $$\left\{ \frac{8 \cdot x - e^x}{4 \cdot y(x)^3 - 1} \right\}$$

- $\text{First\_derivative} := \text{diff}(y(x), x) = \text{op}(d, 1);$

  $$\frac{\partial}{\partial x} y(x) = \frac{8 \cdot x - e^x}{4 \cdot y(x)^3 - 1}$$

- $w := \text{diff}(\text{First\_derivative}, x);$

  $$\frac{\partial^2}{\partial x^2} y(x) = \frac{-e^x + 8}{4 \cdot y(x)^3 - 1} - \frac{12 \cdot y(x)^2 \cdot \frac{\partial}{\partial x} y(x) \cdot (8 \cdot x - e^x)}{(4 \cdot y(x)^3 - 1)^2}$$

- $\text{Second\_derivative} := \text{subs}(w, \text{First\_derivative});$

  $$\frac{\partial^2}{\partial x^2} y(x) = \frac{-e^x + 8}{4 \cdot y(x)^3 - 1} - \frac{12 \cdot y(x)^2 \cdot (8 \cdot x - e^x)^2}{(4 \cdot y(x)^3 - 1)^3}$$

Figure 2.11. A MuPAD implementation of the MPL dialogue in Figure 2.8. (Implementation: MuPAD (mnb).)

logue. The next three prompts, however, correspond to the single statement <3> in the MPL dialogue. At the third prompt, we use MuPAD's subs operator to replace the derivative diff(y(x),x) in the previous expression by the symbol Dy. This step is required because MuPAD's solve operator cannot solve for the expression diff(y(x),x), even though it can solve for other function forms. At the fourth prompt, MuPAD's Solve operator solves the equation v2 for Dy. Notice that we have included the option IgnoreSpecialCases because, without this, the system performs a more detailed analysis of the equation and also includes solutions for which the denominator $4 \cdot y(x)^3 - 1 = 0$. These special solutions are not required in our dialogue. At the fifth prompt, we use MuPAD's op operator (which selects operands of an expression) to extract the solution from the set d and include diff(y(x),x) on the left side of an equation so that First_derivative corresponds to the output of <3> in the MPL dialogue.

The operations at the next two prompts are the same as those at <4> and <5> in the MPL dialogue.

Exercises

For the exercises in this section, the following operators are useful:

- In Maple, the **expand**, **diff**, **subs**, **solve**, **op**, and **dsolve** operators. (Implementation: Maple (mws).)
- In Mathematica, the **Expand**, **D**, **ReplaceAll**, **Solve**, **Part**, and **DSolve** operators. (Implementation: Mathematica (nb).)
- In MuPAD, the **expand**, **diff**, **subs**, **solve**, **op**, and **ode** operators. (Implementation: MuPAD (mnb).)

1. In this exercise we ask you to give an interactive dialogue in a CAS similar to the one in Figure 2.8 that simulates the mathematical discourse in Figure 2.12. Use a CAS's command for solving a differential equation to obtain the general solution as in Expression (2.16), but don't use this command to obtain the arbitrary constant in the solution. Rather, use statements similar to those in Figure 2.8 to set up an equation for the arbitrary constant and solve the equation. The last statement in the dialogue should return an equation similar to Expression (2.19).

Consider the differential equation and initial condition:

$$\frac{dy}{dx} + y = x + \exp(-2x), \quad y(1) = 3. \tag{2.15}$$

The general solution to this equation is given by

$$y = x - 1 - \exp(-2x) + c \exp(-x), \tag{2.16}$$

where c is an arbitrary constant. To find c, we substitute the initial condition $y(1) = 3$ into Equation (2.16) and obtain an equation for c:

$$3 = -e^{-2} + c e^{-1}. \tag{2.17}$$

Solving for c, we obtain
$$c = 3e + e^{-1}. \tag{2.18}$$

Substituting Equation (2.18) into Equation (2.16), we obtain the particular solution to the differential equation:

$$y = x - 1 - \exp(-2x) + (3e + e^{-1}) \exp(-x). \tag{2.19}$$

Figure 2.12. A mathematical discourse that obtains the arbitrary constant in the solution of a differential equation.

2. Consider the second order differential equation

$$\frac{d^2y(x)}{dx^2} + 5\frac{dy(x)}{dx} + 6\,y(x) = \sin(x), \quad y(0) = 2. \quad \frac{dy}{dx}(0) = 1.$$

This equation has a general solution that involves two arbitrary constants which are found by substituting the two initial conditions into both the general solution and its derivative and then solving the resulting system of linear equations.

Give an interactive dialogue in a CAS similar to the one in Figure 2.8, which finds the general solution to the differential equation, sets up the equations for the arbitrary constants, solves for the arbitrary constants, and then substitutes them back into the general solution. Use a CAS's command for solving a differential equation to obtain the general solution to the differential equation, but don't use this command to obtain the arbitrary constants in the solution. Rather, use statements similar to those in Figure 2.8 to obtain the arbitrary constants. The last statement in the dialogue should return an equation equivalent to

$$y(x) = (1/10)\,\sin(x) - (1/10)\,\cos(x) + (36/5)e^{-2\,x} - (51/10)\,e^{-3\,x}.$$

3. (a) Consider the polynomial $y = a\,x^3 + b\,x^2 + c\,x + d$. Give an interactive dialogue in a CAS that finds the coefficients a, b, c, and d such that at $x = 2$,

$$y = 5, \quad \frac{dy}{dx} = -2, \quad \frac{d^2y}{dx^2} = 2, \quad \frac{d^3y}{dx^3} = -3.$$

The last statement should return the polynomial with the numerical values for the coefficients.

(b) Use the dialogue to show there are infinitely may expressions of the form $y = (a\,x + b)/(c\,x + d)$ that satisfy the conditions in part (a).

(c) Use the dialogue to show it is impossible to find an expression of the form $y = (a\,x + b)/(c\,x + d)$ that satisfies the conditions

$$y = 1, \quad \frac{dy}{dx} = 2, \quad \frac{d^2y}{dx^2} = 3, \quad \frac{d^3y}{dx^3} = 4.$$

2.2 Expression Evaluation

The term *expression evaluation* (or just *evaluation*) refers to the actions taken by a CAS in response to an input expression. These actions include:

1. the analysis of the structure of an expression and the translation of this structure into an internal form that is used by the CAS to represent the expression;

2. the evaluation of assigned variables and mathematical operators that appear in an expression; and

3. the application of some elementary algebraic and trigonometric simplification rules.

In this section we consider the evaluation of variables and operators, and take a brief look at the simplification process. Expression structure is described in detail in Chapter 3.

Variable and Operator Evaluation

Figure 2.13 shows an MPL dialogue that gives some examples of variable and operator evaluation. At <1> the expression $t + 1$ is assigned to x and at <2> a polynomial in x is assigned to y. Since x has been assigned, its value is included in the expression for y. In a similar way at <3>, the values for x and y are included in the expression and the *Factor* operator is evaluated. Statements <4> and <5> show that the evaluation process applies to function names as well as other variables in an expression.

But now, what happens when the value of an assigned variable is another expression which also contains assigned variables? Statements <6> through <9> illustrate what can happen in this situation. At <6>, <7>, and <8>, the variables u, v, and w are assigned values where all variables to the right of the assignment symbols are unassigned. What is the value of u^2 after the execution of these assignments? Statement <9> contains two responses that illustrate two different approaches to variable evaluation. In the first approach, called *single-level evaluation*, the value of u is the value it was originally assigned $(v + 2)$, and the assigned value of v in this expression is ignored. In other words, with single-level evaluation only one level of active assignments is used.

In the second approach, called *multi-level evaluation*, the evaluation process uses all active assignments. In this case, the value of u^2 is obtained using three levels of assignments

$$u^2 \to (v + 2)^2 \to (w^2 + 2)^2 \to ((t + 3)^2 + 2)^2. \qquad (2.20)$$

Since some systems provide a way to control the evaluation level, MPL provides the *Evaluate* operator for this purpose. At <10>, we evaluate u^2 using two levels of assignments.

Occasionally, it is useful to suppress the evaluation of a variable or operator. We denote this operation in MPL by placing the variable or operator name in quotes. For example, at <11> we suppress evaluation of the *Derivative* operator and at <12> suppress evaluation of the assigned variable w. Finally, there are times when it is necessary to unassign or

<1> $x := t + 1;$
$$\rightarrow \quad x := t + 1;$$
<2> $y := x^2 + 4 * x + 4;$
$$\rightarrow \quad y := (t+1)^2 + 4(t+1) + 4$$
<3> $x * Factor(y);$
$$\rightarrow \quad (t+1)(t+3)^2$$
<4> $z := f;$
$$\rightarrow \quad f;$$
<5> $z(x);$
$$\rightarrow \quad f(t+1)$$
<6> $u := v + 2;$
$$\rightarrow \quad u := v + 2;$$
<7> $v := w^2;$
$$\rightarrow \quad v := w^2;$$
<8> $w := t + 3;$
$$\rightarrow \quad w := t + 3$$
<9> $u^2;$
$$\rightarrow \quad (v+2)^2 \quad \text{(single-level evaluation)}$$
$$\rightarrow \quad ((t+3)^2 + 2)^2 \quad \text{(multi-level evaluation)}$$
<10> $Evaluate(u^2, 2);$
$$\rightarrow \quad (w^2 + 2)^2$$
<11> $"Derivative"(t^2, t);$
$$\rightarrow \quad \frac{d\,(t^2)}{d\,t}$$
<12> $Derivative("w", t);$
$$\rightarrow \quad 0$$
<13> $Unassign(w);$
$$\rightarrow \quad w$$
<14> $Algebraic\_expand((w+1) * (w+2));$
$$\rightarrow \quad w^2 + 3\,w + 2$$

Figure 2.13. An MPL dialogue that shows examples of variable and operator evaluation. (Implementation: Maple (mws), Mathematica (nb), MuPAD (mnb).)

MPL	Maple	Mathematica	MuPAD
multi-level (dialogues), single-level (in procedures)	multi-level (interactive mode), single-level (in procedures)	multi-level in interactive mode and procedures	multi-level (interactive mode), single-level (in procedures)
control evaluation level $Evaluate(u, n)$	eval(u,n)	not available	level(u,n)
suppress evaluation $"u"$ $"Derivative"(u, x)$	'u' 'diff'(u,x)	HoldForm[u] HoldForm[D][u,x], and release suppressed evaluation with ReleaseHold[u]	hold(u) hold(diff)(u,x)
$Unassign(u)$	unassign('u')	u = .	delete(u)

Figure 2.14. Evaluation concepts in Maple, Mathematica, and MuPAD.

remove the value of an assigned variable. In MPL, the *Unassign* operator is used for this purpose. At <13>, we apply *Unassign* to w, which means at <14> w acts as a symbol in the mathematical expression.

In Figure 2.14, we summarize the evaluation concepts considered above in the Maple, Mathematica, and MuPAD systems. Observe that all three systems use multi-level evaluation in the interactive mode, while both Maple and MuPAD switch to single-level evaluation inside procedures[7]. In MPL, we also use multi-level evaluation in dialogues, and following Maple and MuPAD, use single level evaluation in procedures.

Automatic Simplification

The term *automatic* (or *default*) *simplification* refers to the mathematical simplification rules that are applied to an expression during the evaluation process. In computer algebra systems, this usually involves the "obvious" simplification rules from algebra and trigonometry that remove extraneous symbols from an expression and transform it to a standard form.

---

[7]Procedures in a CAS language are like procedures or functions in a conventional programming language. A procedure in the Maple language is given in Figure 1.3 on page 6. We consider procedures in Chapter 4.

<1> $2 + 3/4 + 5/6;$

$$\rightarrow \quad \frac{43}{12}$$

<2> $x + y + 2 * x;$

$$\rightarrow \quad 3x + y$$

<3> $x * y * x^2;$

$$\rightarrow \quad x^3 y$$

<4> $1 * x^3 + a * x^0 + b * x^1 + 0 * x^2;$

$$\rightarrow \quad a + b\,x + x^3$$

<5> $x * y + 3 * y * x;$

$$\rightarrow \quad 4\,x\,y$$

<6> $\sin(\pi/2);$

$$\rightarrow \quad 1$$

<7> $\ln(e^2);$

$$\rightarrow \quad 2$$

<8> $\arctan(1);$

$$\rightarrow \quad \pi/4$$

<9> $i^2;$

$$\rightarrow \quad -1$$

<10> $e^{(-i*\pi)};$

$$\rightarrow \quad -1$$

<11> $0 \leq 1$ **and** $1 \leq 2;$

$$\rightarrow \quad \textbf{true}$$

<12> P **and** P **and** $Q;$

$$\rightarrow \quad P \textbf{ and } Q$$

Figure 2.15. An MPL dialogue that shows some examples of automatic simplification. (Implementation: Maple (mws), Mathematica (nb), MuPAD (mnb).)

The MPL dialogue in Figure 2.15 illustrates some of these obvious simplifications. Example <1> shows a simplification that involves the sum of rational numbers. Example <2> shows that automatic simplification combines numerical coefficients of like terms. The next example <3> illustrates a similar simplification in which integer exponents of the common base x are combined. Example <4> illustrates some simplification rules that involve the integers 0 and 1. Notice that after evaluation, the x^3 term appears at the right end of the expression. This reordering, which is an application of the commutative law of addition, serves to put the result in a more readable form and, in some cases, contributes to the simplification process[8]. The next example <5> illustrates this point. To simplify this expression, the term $3 * y * x$ is first reordered (using the commutative law for multiplication) to $3 * x * y$ after which the coefficients of the two like terms are combined. Examples <6>, <7>, and <8> illustrate automatic simplification rules that involve known functions, while Examples <9> and <10> illustrate simplification rules that involve reserved symbols.

Examples <11> and <12> illustrate the automatic simplification rules that are applied in some systems to logical expressions as data objects[9]. In Example <12>, P and Q are unassigned identifiers and the simplification follows from the general logical rule P **and** $P \rightarrow P$.

The examples in Figure 2.15 are roughly similar to what happens in a real computer algebra system. However, since there is no consensus about which simplification rules should be included in automatic simplification, the process can vary somewhat from system to system.

Figure 2.16 shows an interactive dialogue with the Macsyma system that shows what happens when automatic simplification is suppressed. At the prompt (c1) we assign an expression to u and at (c2) turn off the automatic simplifier by assigning the value **false** to the variable **simp**. At (c3) we differentiate u and obtain an expression that is so involved it is difficult to interpret[10]. At (c4) we turn the automatic simplifier back on and at (c5) obtain a much more reasonable form for the derivative.

In MPL (as in a CAS), all expressions in dialogues and computer programs operate in the context of automatic simplification. This means:

- All input operands to mathematical operators are automatically simplified before the operators are applied.

- The result obtained by evaluating an expression is in automatically simplified form.

Since automatic simplification is so central to the programming process, it is a good idea to understand which simplification rules are applied by the process and which are not. For now, the exercises in this section can be used to explore the automatic simplification process in a CAS.

---

[8]The reordering process in Mathematica and MuPAD is similar to what is described here. The reordering process in Maple is handled in a different way (see Cohen [24], Section 3.1).

[9]Maple obtains <11> and <12>. Mathematica obtains <11>, but not <12>. MuPAD obtains <12>, but not <11>.

[10]Notice that Macsyma uses logarithmic differentiation to differentiate e^{x^2}. Logarithmic differentiation provides a way to differentiate general powers of the form $f(x)^{g(x)}$.

(c1)   u : a*x + x*exp(x∧2);

(d1)
$$x\,e^{x^2} + a\,x$$

(c2)   simp : false;

(d2)
false

(c3)   diff(u,x);

(d3)
$$1\,a + 0\,x + e^{x^2}\left(e^{-1}x^2\,0 + \log(e)\,(2x)\right)x + 1\,e^{x^2}$$

(c4)   simp : true;

(d4)
true

(c5)   diff(u,x);

(d5)
$$2\,x^2\,e^{x^2} + e^{x2} + a$$

Figure 2.16. An interactive dialogue with the Macsyma system that shows what happens when automatic simplification is suppressed.

In Chapter 3, we show how automatic simplification modifies the structure of expressions, which in turn leads to simpler algorithms and programs. In Cohen [24], Chapter 3, we give the formal algebraic properties of automatically simplified expressions and describe an algorithm that obtains the simplified form.

Exercises

1.   (a) Consider the following transformations of powers[11]:
     i. $x^2\,x^3 \to x^5$.
     ii. $x^{1/2}\,x^{1/3} \to x^{5/6}$.
     iii. $x^a\,x^b \to x^{a+b}$.

---

[11]Some of the power transformations in this problem are only valid in certain (real or complex) contexts.

iv. $(x^2)^3 \to x^6$.

v. $(x^a)^2 \to x^{2a}$.

vi. $(x^2)^{1/2} \to |x|$.

vii. $(x^{1/2})^2 \to x$.

viii. $(x^2)^a \to x^{2a}$.

ix. $(x\,y)^2 \to x^2\,y^2$.

x. $(x\,y)^{1/3} \to x^{1/3}\,y^{1/3}$.

xi. $(x\,y)^a \to x^a\,y^a$.

Which of these transformations is obtained with automatic simplification?

(b) Based on the data obtained in part (a), give a summary of how the power translations are applied in automatic simplification.

2. The algebraic operations addition and multiplication obey the following distributive laws:

$$a \cdot (b + c) = a \cdot b + a \cdot c, \qquad (a + b) \cdot c = a \cdot c + b \cdot c$$

(a) Consider the following transformations which are based on these laws:

i. $2\,x + 3\,x \to 5\,x$.

ii. $(1 + x) + 2\,(1 + x) \to 3\,(1 + x)$.

iii. $2\,x + \sqrt{2}\,x \to (2 + \sqrt{2})\,x$.

iv. $a\,x + b\,x \to (a + b)\,x$.

v. $(a + b)\,x \to a\,x + b\,x$.

vi. $2\,(x + y) \to 2\,x + 2\,y$.

vii. $-(x + y) \to -x - y$.

viii. $a\,(x + y) \to a\,x + a\,y$.

Which of the these transformations are obtained with automatic simplification?

(b) Based on the data obtained in part (a), give a summary of how the distributive laws are applied in automatic simplification.

3. (a) Consider the following transformations of the sin function:

i. $\sin(0) \to 0$.

ii. $\sin(\pi/2) \to 1$.

iii. $\sin(\pi/5) \to \dfrac{\sqrt{2}\sqrt{5 - \sqrt{(5)}}}{4}$.

iv. $\sin(\pi/60) \to \dfrac{\sqrt{5+\sqrt{5}}}{8} - \dfrac{\sqrt{5+\sqrt{5}}\sqrt{3}}{8} - \dfrac{\left(-\frac{\sqrt{5}}{4}+1/4\right)\sqrt{2}}{4}\dfrac{\left(-\frac{\sqrt{5}}{4}+1/4\right)\sqrt{2}\sqrt{3}}{4}$.

v. $\sin(15\,\pi/16) \to \sin(\pi/16)$.

vi. $\sin(-x) \to -\sin(x)$.

vii. $\sin(-x + 1) \to -\sin(x - 1)$.

  viii. $\sin(x + \pi/2) \rightarrow \cos(x)$.
   ix. $\sin(x + 2\pi) \rightarrow \sin(x)$.
    x. $\sin(a + b) \rightarrow \sin(a)\cos(b) + \cos(a)\sin(b)$.
   xi. $\sin(a)\cos(b) + \cos(a)\sin(b) \rightarrow \sin(a + b)$.
  xii. $\sin^2(x) + \cos^2(x) \rightarrow 1$.

  Which of the these transformations is obtained with automatic simplification?

  (b) Based on the data obtained in part (a), give a summary of the trans-
      formation rules for the sin function which are obtained with automatic
      simplification.

4. In this problem we ask you to explore how the indeterminate forms $0/0$
   and 0^0 are handled in automatic simplification.

   (a) Enter each of the following expressions in the interactive mode of a
       CAS.

       i. $0/0$.
      ii. 0^0.
     iii. $(a(x + y) - ax - ay)/(x - x)$.
      iv. $(x - x)/(a(x + y) - ax - ay)$.
       v. $(x - x)^{a(x+y)-ax-ay}$.
      vi. $(a(x + y) - ax - ay)^{x-x}$.

   (b) Based on the data obtained in part (a), give a summary of how inde-
       terminate forms are handled by automatic simplification.

5. Enter each of the following expressions in the interactive mode of a CAS:

$$-b, \qquad a - 2 * b, \qquad 1/a^2, \qquad a/b.$$

   Although each of the expressions is returned in the form it was entered,
   some "hidden" transformations have been applied. In other words, the
   internal form used by the CAS is different from the displayed form. Use
   the operand selection operator in a CAS to determine the internal form.
   (Use op in Maple and MuPAD, and Part in Mathematica.)

6. This exercise refers to the Macsyma dialogue in Figure 2.16 on page 55.
   What simplification rules are used to obtain (d5) instead of (d3)?

7. Consider a CAS such as Mathematica or MuPAD where terms in a sum
   or factors in a product are reordered as part of automatic simplification.
   Experiment with the CAS to determine how it carries out the ordering
   process. Try polynomials such as <4> in Figure 2.15 as well as more involved expressions. For example, are any terms or factors in

$$(1 + zy^2 + (a + 1)b + c)(a + 1)$$

   reordered by automatic simplification?

2.3   Mathematical Programs

Simply put, an MPL *mathematical program* (or *mathematical algorithm*) is a sequence of statements in the MPL language that can be implemented in terms of the operations and control structures available in a computer algebra programming language. The design and implementation of mathematical programs is a major theme of this book.

In a sense, the MPL dialogue in Figure 2.8 is an example of a simple interactive program. What we really have in mind, however, are more involved programs that have the following features:

1. The statements in the program are viewed collectively as a unit which either is entered at a single prompt or input region in the interactive mode or, for larger programs, is contained in a text file that is loaded into the system.

2. The program statements include mathematical expressions, assignment statements, decision statements, iteration statements, and function and procedure definitions[12].

3. As with conventional programs, some statements serve as input statements, some statements are for intermediate calculations for which the output is not displayed, and some statements serve as output statements that display the result of a computation.

4. The program is designed in a general way so that it performs a calculation for a class of problems rather than for a single problem.

For an example of a program that incorporates some of these points, let's consider again the computation of the first and second derivatives of an implicitly defined function such as

$$\exp(x) + y^3 = 4\,x^2 + y.$$

This problem, which was considered in Section 2.1, involves the manipulations in Equations (2.8) through (2.12), and an **MPL** dialogue that performs the calculations is given in Figrue 2.8 on page 42. This dialogue assumes that x is the independent variable, y is the dependent variable, and requires that y be expressed as the function form $y(x)$.

In this section we modify the program to permit a choice of mathematical variable names and do not require that the dependent variable be

---

[12]Decision statements, iteration statements, and function and procedure definitions are described in Chapter 4.

```
1    u_in := exp(s) + t⁴ = 4 * s² + t :
2    x_var := s :
3    y_var := t :
4    u_new := Substitute(u_in, y_var = y_var(x_var)) :
5    u_p := Derivative(u_new, x_var):
6    First_derivative := Solve(u_p, Derivative(y_var(x_var), x_var));
7    u_pp := Derivative(First_derivative, x_var) :
8    Second_derivative := Substitute(u_pp, First_derivative);
```

Figure 2.17. An MPL mathematical program that obtains the first and second derivatives of an implicit function. (Implementation: Maple (mws), Mathematica (nb), MuPAD (mnb).)

expressed as a function form. By simply modifying the input statements, we can obtain

$$\frac{dy}{dx}, \quad \frac{d^2y}{dx^2} \quad \text{or} \quad \frac{dx}{dy}, \quad \frac{d^2x}{dy^2},$$

or, for that matter, if the input expression is expressed in terms of the variables s and t as $\exp(s) + t^3 = 4\,s^2 + t$, the derivatives

$$\frac{dt}{ds}, \quad \frac{d^2t}{ds^2} \quad \text{or} \quad \frac{ds}{dt}, \quad \frac{d^2s}{dt^2}.$$

An MPL program that performs these calculations is given in Figure 2.17. Observe that some statements are terminated by a colon (lines 1, 2, 3, 4, 5, and 7) and some by a semicolon (lines 6 and 8). This notation is interpreted as follows: statements that end with a colon suppress the display of the output, while those that end with a semicolon display the output. Most computer algebra systems allow control of output display, although the termination symbols vary from system to system[13].

Lines 1 through 3 serve as input statements for the program. Since the program is designed to allow a choice of mathematical variable names, we have chosen programming variable names (u_in, x_var, y_var, etc.) that are unlikely to be used as mathematical variables. At line 1 we assign an input expression, and at lines 2 and 3, we initialize two programing variables x_var and y_var which contain the mathematical variables (s and t for

---

[13] In both Maple and MuPAD, statements that are terminated with a colon suppress the output, while those that are terminated with a semicolon display the output. In Mathematica, statements that are terminated with a semicolon suppress the output, while those without a terminating symbol display the output.

this input) which serve as the independent and dependent variables. With these two assignments, the output of the program is the derivatives

$$\frac{dt}{ds} \text{ and } \frac{d^2t}{ds^2}.$$

The derivative operations at lines 5 and 7 require that the dependent variable t be expressed as the function form $t(s)$. Since this is not done in line 1, we account for this at line 4 with a substitution that replaces each t in u_in by $t(s)$. Except for changes in notation, lines 5 through 8 are similar to those in Figure 2.8. With the choice of input, the outputs from lines 6 and 8 are

$$\frac{dt(s)}{ds} = \frac{-\exp(s) + 8\,s}{4\,t(s)^3 - 1}, \tag{2.21}$$

$$\frac{d^2t(s)}{ds^2} = \frac{-\exp(s) + 8}{4\,t(s)^3 - 1} - 12\frac{\left(-\exp(s) + 8\,s\right)^2 t(s)^2}{\left(4\,t(s)^3 - 1\right)^3}. \tag{2.22}$$

Observe that the dependent variable t is expressed in function notation $t(s)$, even though this is not done in the input at line 1. In Exercise 1 we describe a modification of the program that removes this function notation from the output.

Case Study: General Quadratic Equations and Rotation of Axes

We conclude this section with a more involved MPL program that obtains the change of form of a quadratic equation under rotation of coordinate axes.

A general quadratic equation in x and y has the form

$$A\,x^2 + B\,x\,y + C\,y^2 + D\,x + E\,y + F = 0, \tag{2.23}$$

where the coefficients are rational numbers and at least one of the coefficients $A, B, C \neq 0$. This equation represents one of the following eight graphs in the plane:

1. a circle (such as $x^2 + y^2 - 1 = 0$).

2. an ellipse (such as $x^2 + 2\,y^2 - 1 = 0$).

3. a single point (such as $x^2 + y^2 = 0$ or $(x, y) = (0, 0)$).

4. an empty graph (such as $x^2 + y^2 = -1$).

5. a hyperbola (such as $x^2 - y^2 = 1$).

6. a parabola (such as $x^2 - y = 0$).

7. two intersecting lines (such as $x^2 - y^2 = 0$ or $x = \pm y$).

8. a single line (such as $x^2 + 2xy + y^2 = (x + y)^2 = 0$ or $x = -y$).

If $B = 0$ in Equation (2.23), it is a simple matter to determine the type of graph and some of its important features by using the techniques of analytical geometry. If $B \neq 0$, the analysis is more involved. However, by rotating the coordinate system, it is possible to transform Equation (2.23) into a general quadratic equation in terms of new variables (u, v) so that the coefficient of the uv term is zero.

Consider the coordinate rotation shown in Figure 2.18, where the (u, v) coordinate system is rotated by an angle α from the (x, y) system. To find a relationship between the (x, y) and (u, v) coordinates, we have

$$x = r \cos(\alpha + \beta) \quad y = r \sin(\alpha + \beta) \tag{2.24}$$

and

$$u = r \cos(\beta) \quad v = r \sin(\beta). \tag{2.25}$$

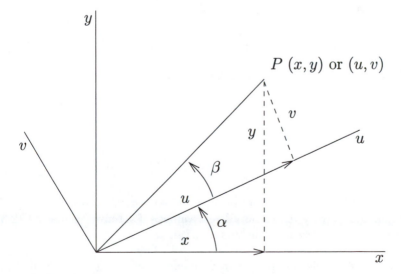

Figure 2.18. The point P in the (x, y) and (u, v) coordinate systems.

By expanding the trigonometric expressions in Equations (2.24), we obtain

$$\begin{aligned} x &= r\cos(\alpha)\cos(\beta) - r\sin(\alpha)\sin(\beta), & (2.26) \\ y &= r\sin(\alpha)\cos(\beta) + r\cos(\alpha)\sin(\beta). & (2.27) \end{aligned}$$

Substituting Equations (2.25) into Equation (2.27), we obtain the coordinate transformation

$$x = u\cos(\alpha) - v\sin(\alpha), \quad y = u\sin(\alpha) + v\cos(\alpha). \qquad (2.28)$$

By substituting Equations (2.28) into the original Equation (2.23), we obtain a quadratic equation in the (u, v) system

$$A'\,u^2 + B'\,u\,v + C'\,v^2 + D'\,u + E'\,v + F' = 0, \qquad (2.29)$$

where

$$\begin{aligned} A' &= A\cos^2(\alpha) + B\cos(\alpha)\sin(\alpha) + C\sin^2(\alpha), \\ B' &= B\left(\cos^2(\alpha) - \sin^2(\alpha)\right) + 2\left(C - A\right)\sin(\alpha)\cos(\alpha), \\ C' &= A\sin^2(\alpha) - B\sin(\alpha)\cos(\alpha) + C\cos^2(\alpha), \qquad (2.30) \\ D' &= D\cos(\alpha) + E\sin(\alpha), \\ E' &= -D\sin(\alpha) + E\cos(\alpha), \\ F' &= F. \end{aligned}$$

To find an α so that the coefficient of the $u\,v$ term $B' = 0$, we use the trigonometric reduction rules

$$2\sin(\alpha)\cos(\alpha) = \sin(2\,\alpha), \quad \cos^2(\alpha) - \sin^2(\alpha) = \cos(2\,\alpha),$$

so that the second equation in (2.30) becomes

$$B' = B\cos(2\,\alpha) + (C - A)\sin(2\,\alpha).$$

Setting $B' = 0$, we obtain when $B \neq 0$

$$\cot(2\,\alpha) = \frac{A - C}{B}. \qquad (2.31)$$

(When $B = 0$, a rotation is not needed and so $\alpha = 0$.) When $0 < \alpha < \pi/2$, the function $\cot(2\,\alpha)$ takes on all real values. Therefore, Equation (2.31) defines a unique rotation α in this interval.

Example 2.2. Consider the quadratic equation $x^2 + 2\,x\,y + y^2 = 0$. Since the left side can be factored as $(x + y)^2$, the quadratic equation represents the straight line $y = -x$. To find the equation of the line in the (u, v) system, we have from Equation (2.31)

$$\cot(2\,\alpha) = \frac{A - C}{B} = \frac{1 - 1}{2} = 0$$

and so $\alpha = \pi/4$. Using Equation (2.30), we obtain the simple equation $u = 0$ for the line in the new coordinate system. □

To apply Equation (2.30) in more involved situations, we need expressions for $\sin(\alpha)$ and $\cos(\alpha)$. Since $0 < 2\,\alpha < \pi$, $\cos(2\,\alpha)$ has the same sign as $\cot(2\,\alpha)$. Therefore Equation (2.31) implies

$$\cos(2\,\alpha) = \frac{B}{|B|}\frac{A - C}{\sqrt{(A - C)^2 + B^2}} \tag{2.32}$$

where $B/|B|$ is included so that $\cos(2\,\alpha)$ has the correct sign. We obtain values for $\sin(\alpha)$ and $\cos(\alpha)$ with Equation (2.32) and the identities

$$\cos(\alpha) = \sqrt{\frac{1 + \cos(2\,\alpha)}{2}}, \qquad \sin(\alpha) = \sqrt{\frac{1 - \cos(2\,\alpha)}{2}}.$$

By using these identities together with Equation (2.32), we obtain the coefficients in Equation (2.30) in the new system without finding α.

Before giving a program that performs these calculations, we describe two polynomial operators that are available in most computer algebra systems. Consider the polynomial in x

$$w = w_n x^n + w_{n-1} x^{n-1} + \cdots + w_0, \tag{2.33}$$

where the *coefficients* w_j can be integers, fractions, symbols, or even more involved expressions, x represents a variable or a more involved expression, and n is a non-negative integer. Recall, the largest power of x in a polynomial is called the *degree* of the polynomial.

The two most important operators for polynomials are the *Degree* and *Coefficient* operators[14]. The operator *Degree*(w, x) returns the degree of the polynomial w with respect to an expression x.

Example 2.3.

$$Degree(3\,x^2 + x + 5, x) \rightarrow 2,$$

$$Degree(a\,\sin^2(x) + b\,\sin(x) + c,\ \sin(x)) \rightarrow 2.$$

---

[14] The *Degree* and *Coefficient* operators are described in greater detail in Chapter 6. In Chapter 6, these operators are called *Degree_gpe* and *Coefficient_gpe*, where the suffix *gpe* stands for general polynomial expression (see Definition 6.14 on page 223).

Observe that in the second example the expression is considered a polynomial in the function form $\sin(x)$. (Implementation: Maple (mws), Mathematica (nb) MuPAD (mnb).) $\square$

The operator $Coefficient(w, x, j)$ returns the coefficient w_j of x^j in Equation (2.33).

Example 2.4.

$$Coefficient(3\, x^2 + x + 5,\ x,\ 2) \rightarrow 3,$$

$$Coefficient(a\, \sin^2(x) + b\, \sin(x) + c,\ \sin(x),\ 2) \rightarrow a,$$

$$Coefficient(a\, x + b\, x + c,\ x,\ 1) \rightarrow a + b,$$

$$Coefficient(Coefficient(2\, x^2 + 3\, x\, y + 4\, y^2 + 5\, x + 6\, y + 7, x, 1),\ y,\ 0) \rightarrow 5.$$

In the last example, the inner application of the $Coefficient$ operator obtains $3\, y + 5$ and the outer application obtains the value 5. (Implementation: Maple (mws), Mathematica (nb), MuPAD (mnb).) $\square$

The degree and coefficient operations in Maple, Mathematica and Mu-PAD are given in Figure 2.4 on page 35.

An MPL program that obtains an expression for the polynomial in the (u, v) coordinate system is given in Figure 2.19. The input to the program is given in the assignment at line 1. Notice that we permit both sides of the equation to contain terms of the polynomial and combine the two sides using the $Operand$ selection operator. The output of the program is the equation obtained by evaluating the expression in line 17. For the expression in line 1, the output is the equation $(1/2)\, u^2 - (1/2)\, v^2 - 1 = 0$.

The programs considered so far are quite elementary. In later chapters we introduce other MPL mathematical operators and language features that enable us to construct more involved and interesting programs.

Exercises

For the exercises in this section, the following operators are useful:

- In Maple, the `coeff`, `expand`, `abs`, `diff`, `subs`, `solve`, `op`, `int`, and `dsolve` operators. (Implementation: Maple (mws).)

- In Mathematica, the `Coefficient`, `Expand`, `Abs`, `D`, `Derivative`, `ReplaceAll`, `Solve`, `Part`, `Integral`, and `DSolve` operators. (Implementation: Mathematica (nb).)

- In MuPAD, the `coeff`, `expand`, `abs`, `diff`, `subs`, `solve`, `op`, `int`, and `ode` operators (Implementation: MuPAD (mnb).)

```
1      eq := x * y = 1 :
2      w := Operand(eq, 1) − Operand(eq, 2) :
3      A := Coefficient(w, x, 2) :
4      B := Coefficient(Coefficient(w, x, 1), y, 1) :
5      C := Coefficient(w, y, 2) :
6      D := Coefficient(Coefficient(w, x, 1), y, 0) :
7      E := Coefficient(Coefficient(w, y, 1), x, 0) :
8      F := Coefficient(Coefficient(w, x, 0), y, 0) :
9      g := B/Absolute_value(B) * (A − C)/((A − C)² + B²)^(1/2) :
10     ca := ((1 + g)/2)^(1/2) :
11     sa := ((1 − g)/2)^(1/2) :
12     Ap := A * ca² + B * ca * sa + C * sa² :
13     Cp := A * ca² − B * ca * sa + C * sa² :
14     Dp := D * ca + E * sa :
15     Ep := −D * sa + E * ca :
16     Fp := F :
17     Ap * u² + Cp * v² + Dp * u + Ep * v + Fp = 0;
```

Figure 2.19. An MPL program that transforms a quadratic polynomial in x and y to a quadratic polynomial in u and v without the $u * v$ term. (Implementation: Maple (mws), Mathematica (nb), MuPAD (mnb).)

1. The output of the program in Figure 2.17 is given in Equations (2.21) and (2.22). Observe that the dependent variable t is displayed as a function form. Modify the program so that the output is displayed without function forms including the function forms that appear in the derivative symbols.

2. An implicit equation $f(x, y) = K$ defines a family of curves that depends on the parameter K. We define two families $f(x, y) = K$ and $g(x, y) = C$ to be *orthogonal trajectories* if, whenever a curve $f(x, y) = K$ intersects a curve $g(x, y) = C$, the tangent lines (derivatives) to the curves at the point of intersection are perpendicular. For example, the circles $x^2 + y^2 = K$ (with $K > 0$) and the straight lines $y/x = C$ are orthogonal trajectories. Indeed, using implicit differentiation, a circle has a tangent with slope

$$\frac{dy}{dx} = -x/y,$$

while the straight line has slope

$$\frac{dy}{dx} = y/x.$$

Therefore, at a point of intersection, the circle and line have derivatives that are negative reciprocals and so they intersect at a right angle.

For a given family $f(x, y) = K$, we can find the family of curves that is orthogonal to it by solving a differential equation. For example, for the family of parabolas $f(x, y) = y - x^2 = K$, we first differentiate (implicitly) to obtain a differential equation satisfied by the family of curves

$$\frac{dy}{dx} - 2x = 0.$$

Next, we obtain the differential equation for the orthogonal family by replacing the derivative symbol by its negative reciprocal. Therefore, the orthogonal family to the parabolas satisfies the differential equation

$$\frac{dy}{dx} + 1/2\,x = 0.$$

Solving this equation we obtain the orthogonal family $g(x, y) = \log|x| + 2y = C$.

Now let u represent a family of curves in the form $f(x, y) = K$. Give a program that finds the orthogonal family to $f(x, y) = K$. Test your program for the families $x^2 + y^2 = K$, $y - x^2 = K$, and $x^2 - y^2 = K$.

3. Let w be a general quadratic Equation (2.23) in x and y with $A \neq 0$, $C \neq 0$ and $B = 0$. Give a program that completes the square in x and y. For example, the program should transform $2x^2 + 3y^2 - 4x - 12y + 10 = 0$ to $2(x-1)^2 + 3(y-2)^2 - 4 = 0$. (Do not use the *Factor* operator in this program.)

4. A first order linear differential equation has the form

$$\frac{dy}{dx} = p(x)\,y + q(x). \tag{2.34}$$

It is shown in books on differential equations that the general solution to this equation is

$$y = (1/u)\left(\int u\,q\,dx\right) + C\,(1/u), \tag{2.35}$$

where

$$u = \exp\left(-\int p(x)\,dx\right)$$

and C is an arbitrary constant. The arbitrary constant is found by substituting an initial condition $y(x_0) = y_0$ into the general solution and solving for the constant. Give a program that finds the general solution to a linear differential equation and uses an initial condition to find the arbitrary constant. For example, for

$$\frac{dy(x)}{dx} = x\,y(x) + x, \qquad y(1) = 2, \tag{2.36}$$

the solution returned is equivalent to

$$y(x) = -1 + 3\,\exp(-1/2)\,\exp((-1/2)\,x^2).$$

Assume the input to your program is similar to Equations (2.36).

5. A linear differential Equation (2.34) has a general solution of the form

$$y = f(x) + C g(x) \qquad (2.37)$$

where C is an arbitrary constant (see Equation (2.35) above). In this problem, we are given an expression in the form (2.37) and find the first order linear differential equation which has the expression as a general solution. For example, given

$$y = x \ln(x) + C x, \qquad (2.38)$$

we can find the differential equation by first solving Equation (2.38) for the arbitrary constant

$$C = \frac{y - x \ln(x)}{x}. \qquad (2.39)$$

Differentiating Equation (2.38) we obtain

$$\frac{dy}{dx} = 1 + \ln(x) + C,$$

and substituting the value for C in Equation (2.39) into this expression we obtain

$$\frac{dy}{dx} = y/x + 1.$$

Give a program that finds the differential equation using steps similar to the ones in this example. Assume the input expression has the form (2.37) where $g(x) \neq 0$. Test your program for the functions $y = x \ln(x) + C x$, $y = x + C \sin(x)$, and $y = \exp(x) + C \sin(x)$.

6. A linear second order differential equation has the form

$$\frac{d^2 y}{dx^2} = p(x) \frac{dy}{dx} + q(x) y + r(x). \qquad (2.40)$$

This equation has a general solution of the form

$$y = c_1 f(x) + c_2 g(x) + h(x), \qquad (2.41)$$

where c_1 and c_2 are arbitrary constants. In this problem, we are given an expression in the form (2.41) and find a second order linear differential equation which has the expression as a general solution. For example, for

$$y = c_1 x + c_2 x^2 + x^3,$$

we obtain

$$\frac{d^2 y}{dx^2} = (2/x) \frac{dy}{dx} - (2/x) y + 2 x.$$

Give a program that has an expression of the form Equation (2.41) as input and finds the differential equation. The approach is similar to the one used in Exercise 5, except now we must eliminate two arbitrary constants c_1 and c_2 from the second derivative of the input expression. This problem involves the solution a system of two equations for the two unknowns c_1 and c_2, and to guarantee a solution to the system, assume that $f(x) g'(x) - f'(x) g(x)$ is not identically 0.

7. The method of Lagrange multipliers is a technique for finding the maximum and minimum values of a function of several variables when the independent variables are subject to one or more constraints[15]. For example, to find the maximum and minimum values of $f(x, y)$ where x and y are subject to the side relation $g(x, y) = c$, form a new function

$$L(x, y, \lambda) = f(x, y) - \lambda g(x, y),$$

where the variable λ is called the Lagrange multiplier. Next solve the following three simultaneous equations for the unknowns x, y, and λ:

$$\frac{\partial L}{\partial x} = 0, \quad \frac{\partial L}{\partial y} = 0, \quad \frac{\partial L}{\partial \lambda} = 0. \tag{2.42}$$

The maximum and minimum values (if they exist) occur at the points (x, y) that are solutions to this system. For example, if $f = x - x y + 2$ and the side relation is the line $x - 2 y = 1$, then Equations (2.42) have the solution

$$\lambda = 3/4, \quad x = 3/2, \quad y = 1/4,$$

and the maximum value of f occurs at this point. Give a program that sets up the equations in (2.42) and finds their solution.

8. Let w be a general quadratic equation in x and y with $A = -C \neq 0$.

    (a) Show there is a rotation of axes α that gives $A' = 0$ and $C' = 0$.

    (b) Give a program that finds the equation in the (u, v) system defined by the rotation in part (a).

    (c) Use the algorithm in part (b) to find the equation for $x^2 + x y - y^2 - 1 = 0$ in the (u, v) system.

2.4   Sets and Lists

In computer algebra languages both sets and lists are used to represent collections of mathematical expressions. In this section we give the mathematical properties of sets and lists, and describe the operations that are applied to each of them.

---

[15]See Simmons [88] for an elementary discussion of the method.

Sets

In mathematics, a *set* is defined as simply a collection of objects. In MPL, a set is a finite collection of expressions that is surrounded by the braces { and }. For example, the expression

$$\{x + y = 1, \ x - y = 2\}$$

represents a set with two members, the equations $x + y = 1$ and $x - y = 2$.

Following mathematical convention, MPL sets satisfy the two properties:

1. *The contents of a set does not depend on the order of the elements in the set.* This means that $\{u, v\}$ and $\{v, u\}$ are the same set.

2. *The elements of a set are distinct.* In other words, a set cannot contain duplicate elements.

In MPL, sets are used in situations where the order of expressions in the set is not significant. For example, sets are used in the expression

$$Solve(\{2\,x + 4\,y = 3, \ 3\,x - y = 7\}, \ \{x, \ y\}),$$

since the order of both the equations and the variables does not change the result of the operation.

Algebraic Operations On Sets

Let A and B represent sets and let x represent an arbitrary expression. The following operations are defined for sets:

- *Union of Sets*, $A \cup B$. The *union* of sets A and B is a new set that contains all the elements in A or in B or in both sets. For example,

  $$\{a, \ b, \ c, \ d\} \cup \{c, \ d, \ e, \ f\} \to \{a, \ b, \ c, \ d, \ e, \ f\}.$$

- *Intersection of Sets*, $A \cap B$. The *intersection* of sets A and B is a new set that contains all the elements that are in both A and B. For example,
  $$\{a, \ b, \ c, \ d\} \cap \{c, \ d, \ e, \ f\} \to \{c, \ d\}.$$

- *Difference of Sets*, $A \sim B$. The *difference* of sets A and B is a new set that contains all the elements that are in A but not in B. For example,
  $$\{a, \ b, \ c, \ d\} \sim \{c, \ d, \ e, \ f\} \to \{a, \ b\}.$$

MPL	Maple	Mathematica	MuPAD
set notation $\{a,b,c\}$	`{a,b,c}`	`{a,b,c}`	`{a,b,c}`
$\emptyset$	`{ }`	`{ }`	`{ }`
$A \cup B$	`A union B`	`Union[A,B]`	`A union B`
$A \cap B$	`A intersect B`	`Intersection[A,B]`	`A intersect B`
$A \sim B$	`A minus B`	`Complement[A,B]`	`A minus B`
$x \in A$	`member(x, A)`	`MemberQ[x,A]`	`contains(A,x)`

Figure 2.20. Set operations in Maple, Mathematica, and MuPAD. (Implementation: Maple (mws), Mathematica (nb), MuPAD (mnb).)

- *Set membership,* $x \in A$. The expression $x \in A$ evaluates to **true** if x is in A, and otherwise evaluates to **false**. For example,

$$a \in \{a,\ b,\ c,\ d\} \quad \rightarrow \quad \textbf{true},$$
$$e \in \{a,\ b,\ c,\ d\} \quad \rightarrow \quad \textbf{false}.$$

In the course of manipulating sets, we might obtain the *empty set* or the set with no elements. Following mathematical convention, we represent this set with the reserved symbol $\emptyset$. For example, $\{a,b,c\} \cap \{d,e,f\} \rightarrow \emptyset$.

Most computer algebra systems provide sets along with the algebraic operations described above (see Figure 2.20).

Set Operations on Symbols

Some computer algebra systems allow variable symbols as operands of the set operations $\cup$, $\cap$, and $\sim$, and obtain general set identities as either part of the automatic simplification process or as the output of an operator. This facility is illustrated in the MuPAD dialogue in Figure 2.21. At the first three prompts, automatic simplification obtains the identities

$$A \cup A \cup B = A \cup B,$$
$$A \cap \emptyset = \emptyset,$$
$$(A \cap B) \sim (B \cap A) = \emptyset,$$

and at the fourth prompt, the **expand** operator obtains the distributive law for sets

$$A \cap (B \cup C) = (A \cap B) \cup (A \cap C).$$

Similar results are obtained with the Maple system.

- A union A union B;

  $A \cup B$

- A intersect{ };

  $\emptyset$

- (A intersect B) minus (B intersect A);

  $\emptyset$

- expand(A intersect (B union C));

  $A \cap B \cup A \cap C$

Figure 2.21. General set identities in MuPAD (Implementation: Maple (mws), MuPAD (mnb).)

Lists

An MPL *list* is a finite collection of expressions that is surrounded by the brackets [and]. For example, the expression $[y(x) = 3, \ x = 1]$ is a list with two equations. The empty list, which contains no expressions, is represented by [ ].

Lists are distinguished from sets by the following two properties:

1. *The order of expressions in a list is significant.* This means the expressions $[y(x) = 3, \ x = 1]$ and $[x = 1, \ y(x) = 3]$ represent different lists.

2. *Duplicate elements are permitted in a list.* This means the expressions $[x, y]$ and $[x, y, y]$ represent different lists.

In MPL, lists are used in situations where the order or duplication of expressions is significant. For example, Figure 2.22 illustrates the effect of order on the sequential substitution operation. The *Sequential_substitute* operator shown in the dialogue performs a sequence of substitutions. Since the outcome of this operation depends on the order of substitutions, a list is used to indicate this order. In <1>, the substitution $y(x) = 3$ occurs before $x = 2$ while in <2> this order is reversed. Multiple substitutions, including the *Sequential_substitute* operator, are described in greater detail in Section 3.3.

<1> $Sequential\_substitute(y(x) = m * x + b, \ [y(x) = 3, \ x = 2]);$

$$\rightarrow \quad 3 = 2\,m + b$$

<2> $Sequential\_substitute(y(x) = m * x + b, \ [x = 2, \ y(x) = 3]);$

$$\rightarrow \quad y(2) = 2\,m + b$$

Figure 2.22. An MPL dialogue that illustrates the effect of order in a list on the *Sequential_substitute* operation. (Implementation: Maple (mws), Mathematica (nb), MuPAD (mnb).)

Primitive Operations on Lists

Let L, M, and N represent lists and let x represent an arbitrary expression. The MPL operations for lists reflect the order preserving property:

- *First*(L). If L contains one or more expressions, the operator returns the first expression in L. If $L = [\ ]$, the operator returns the symbol **Undefined**. For example,

$$First([a, b, c]) \rightarrow a.$$

- *Rest*(L). If L contains one or more expressions, the operator returns a new list that contains all expressions in L except the first expression. The original list L is not changed by this operation. If $L = [\ ]$, the operator returns the symbol **Undefined**. For example,

$$Rest([a, b, c]) \rightarrow [b, c].$$

- *Adjoin*(x, L). The operator returns a new list that contains the expression x followed by expressions in L. The original list L is not changed by this operation. For example,

$$Adjoin(d, [a, b, c]) \rightarrow [d, a, b, c].$$

- *Join*$(L, M, \dots, N)$. The operator returns a new list that contains the expressions in the list L followed by the expressions in M and so on. For example,

$$Join([a, b], [b, c], [c, d, e]) \rightarrow [a, b, b, c, c, d, e].$$

MPL	Maple	Mathematica	MuPAD
list notation $[a, b, c]$	[a,b,c]	{a,b,c}	[a,b,c]
empty list $[\,]$	[]	{}	[]
$First(L)$	op(1,L)	First[L]	op(L,1)
$Rest(L)$	[op(2..nops(L),L)]	Rest[L]	[op(L,2..nops(L)]
$Adjoin(x, L)$	[x,op(L)]	Prepend[L,x]	append(L,x)
$Join(L, M)$	[op(L),op(M)]	Join[L,M]	_concat(L,M)
$Reverse(L)$	see Fig. 2.24	Reverse[L]	see Fig. 2.25
$Delete(x, L)$	see Fig. 2.24	Delete[L, Position[L,x]]	listlib :: setDifference(L,[x])
$x \in L$	member(x,L)	MemberQ[x,L]	contains(L,x)

Figure 2.23. List operations in Maple, Mathematica, and MuPAD. (Implementation: Maple (mws), Mathematica (nb), MuPAD (mnb).)

- *Reverse(L)*. The operator returns a new list with elements of the list L in reverse order. The original list L is not changed by this operation. For example,

$$Reverse([a, b, c]) \rightarrow [c, b, a].$$

- *Delete(x, L)*. This operator returns a new list with all instances of x removed from L. The original list L is not changed by this operation. For example,

$$Delete(b, [a, b, c, b]) \rightarrow [a, c].$$

- *List membership, $x \in L$*. The operator returns **true** if x is in L, and otherwise returns **false**. For example,

$$b \in [a, b, c] \rightarrow \textbf{true}.$$

Most computer algebra languages provide lists and most of the list operations described above (see Figure 2.23). Although Maple does not provide the *Reverse* and *Delete* operators, and MuPAD does not provide the *Reverse* operator, these operations can be defined with procedures (see Figures 2.24 and 2.25).

Exercises

1. Let $A = \{a, b, c, d\}$, $B = \{b, d, e, f\}$, and $C = \{a, c, e, f\}$.

   (a) Evaluate
       i. $A \cup B$.
       ii. $A \cap B \cap C$.

```
Reverse:=proc(L)
#Input
#  L:  a list
#Output
#  a new list with the elements of L in reverse order
if L = [ ] or nops(L)=1 then RETURN(L)
else RETURN([op(Reverse([op(2..nops(L),L)])),op(1,L)])
fi
end:

Delete:=proc(x,L)
#Input
#  L:  a list
#Output
#  a new list with all instances of x removed from L
local position;
if member(x,L,position) then
  RETURN([op(1..position-1,L),
          op(Delete(x,[op(position+1..nops(L),L)]))])
else RETURN(L)
fi
end:
```

Figure 2.24. Maple procedures for Reverse and Delete. (Implementation: Maple (txt).)

       iii. $(A \cup B) \cap C$.

       iv. $(A \cup B) \sim C$.

       v. $d \in A$.

  (b) Implement each of the operations in part (a) with a CAS.

2. Let $L = [a, b, c, d]$, $M = [b, d, e, f]$ and $N = [a, c, e, f]$.

  (a) Evaluate

      i. $Rest(Join(L, M, N))$.

      ii. $Adjoin(First(L), M)$.

      iii. $Join(Delete(a, L), Reverse(N))$.

  (b) Implement each of the operations in part (a) with a CAS.

```
Reverse := proc(L)
/*Input
  L:  a list
Output
  a new list with the elements of L in reverse order */
begin
if L = [ ] or nops(L)=1 then return(L)
else return([op(Reverse([op(L,2..nops(L))])),op(L,1)])
end_if
end_proc:
```

Figure 2.25. A MuPAD procedure for `Reverse`. (Implementation: MuPAD (txt).)

3. Let $M = [a, b, c, d]$.

    (a) Give a sequence of MPL statements that performs each of the following operations:

        i. Obtain the last element of M.

        ii. Form a new list with the expression e added to the end of M.

        iii. Form a new list with the second expression removed from M.

    (b) Implement each of the operations in part (a) with a CAS.

Further Reading

2.2 Expression Evaluation. The evaluation process in computer algebra systems is described in Fateman [37]. Evaluation in Maple is described in Heal, Hansen, and Rickard [44]. Evaluation in Mathematica is described in Wolfram [102]. Evaluation in MuPAD is described in Gerhard et al. [40].

2.3 Mathematical Programs. Programming in Maple is described in Monagan et al. [69]. Programming in Mathematica is described in Wolfram [102]. Programming in MuPAD is described in Gerhard et al. [40].

2.4 Sets and Lists. Sets and set operations are described in Maurer and Ralston [65].

3

Recursive Structure of Mathematical Expressions

This chapter is concerned with the structure of mathematical expressions. Since mathematical expressions are the data objects in computer algebra, an understanding of this structure is essential for computer algebra programming.

In Section 3.1 we introduce the concept of recursion and describe a number of ways it is used in mathematics and mathematical algorithms. In this chapter, recursion's main role is to describe the structure of expressions. Since recursion is also an essential programming technique in computer algebra, this topic is covered in detail in Chapter 5.

In Section 3.2 we describe two structural forms for mathematical expressions that correspond to the internal forms used by computer algebra systems before and after automatic simplification. In addition, we introduce four primitive operators that provide a way to analyze and construct expressions. Finally, in Section 3.3 we describe a number of operators, including the *Free_of* and *Substitute* operators, for which the actions depend primarily on the structure of an expression.

3.1 Recursive Definitions and Algorithms

In mathematics, a *recursive* definition or algorithm is one that is defined in terms of a simpler version of itself or sometimes in terms of just another version of itself. The recursion concept is fundamental to nearly all of computer algebra. Indeed, recursiveness in one form or another plays a crucial role in the implementation of many standard operations in computer algebra including simplification, substitution, factorization, solution of equations, differentiation, and integration. In this section, we give a brief introduction to recursion, and explain why it plays such an important role in the manipulation of expressions.

For a simple example, let's consider the operation $n!$ which we first define in a non-recursive way:

$$n! = \begin{cases} 1 & \text{if } n = 0, \\ 1 \cdot 2 \cdots (n-1) \cdot n & \text{if } n > 0. \end{cases} \qquad (3.1)$$

For $n \geq 1$, the factorial operation is defined as the product of all integers from 1 to n. This description does not apply for $n = 0$. Instead, we define $0! = 1$. This convention is a convenient and consistent one for many applications that involve the factorial operation[1].

As a consequence of the definition (3.1), for $n > 0$, $n! = n \cdot (n-1)!$. This relationship forms the basis for a recursive definition of the factorial operation:

$$n! = \begin{cases} 1 & \text{if } n = 0, \\ n \cdot (n-1)! & \text{if } n > 0. \end{cases} \qquad (3.2)$$

This definition is recursive since for $n > 0$, $n!$ is defined in terms of a simpler factorial $(n-1)!$. In this case the adjective *simpler* refers to the factorial operation for a smaller integer value.

The approach in (3.2) is more than just another way to define the factorial operation; it actually suggests another way to implement the calculation. To see what we mean by this, consider first a computation based on the non-recursive definition (3.1). An MPL procedure that performs this calculation is given in Figure 3.1.

The procedure is expressed in the MPL notation and terminology that is used throughout the book to describe mathematical algorithms. Although we will have much to say about this aspect of our pseudo-language in Chapter 4, the examples in this section are simple enough to be understood without a detailed description of the language.

Here is a brief description of the terminology we use in the procedure. A procedure definition in MPL is similar to a function definition in a conventional programming language. The procedure declaration at the top of Figure 3.1 gives the name *Iter_fact* to the sequence of statements in lines 1 through 7. The procedure can be invoked by a statement such as

$$Iter\_fact(4) \rightarrow 24.$$

---

[1] For example, by defining $0! = 1$, the binomial theorem can be expressed in the compact form

$$(x + y)^n = \sum_{i=0}^{n} \frac{n!}{i! \, (n-i)!} x^{n-i} y^i.$$

```
        Procedure   Iter_fact(n);
        Input
            n : non-negative integer;
        Output
            n!;
        Local Variables
            f, i;
        Begin
1           if  n = 0 then
2               Return(1)
3           else
4               f := 1;
5               for  i = 1 to  n do
6                   f := f * i;
7               Return(f)
        End
```

Figure 3.1. An MPL iterative procedure for $n!$. (Implementation: Maple (txt), Mathematica (txt), MuPAD (txt).)

Communication with the procedure is through the input parameter (n in this case) and the *Return* statements in lines 2 and 7. The **if-then-else** statement provides a way to select the appropriate course of action as required by the definition (3.1), and the **for** statement provides a loop that performs the computation. Since this procedure is based primarily on this looping process, it is called an *iterative* procedure.

Let's compare the iterative procedure to a factorial procedure based on the recursive definition (3.2). First, observe how a numerical computation based on (3.2) proceeds:

$$
\begin{aligned}
4! = 4(3!) = 4(3(2!)) = 4(3(2(1!))) &= 4(3(2(1(0!)))) \\
&= 4(3(2(1(1)))) \qquad (3.3) \\
&= 24.
\end{aligned}
$$

To perform the calculation, we repeatedly apply the definition (3.2) until the case $n = 0$ is encountered. Once this point is reached, the value 0! is replaced by the value 1, and the numerical computation proceeds as indicated by the parentheses in the second line of Equation (3.3).

Although this computation has an iterative ring to it, we can give an MPL procedure that is a direct translation of the recursive definition which does not utilize an explicit iteration statement (see Figure 3.2). For $n > 0$,

---

Procedure *Rec_fact*(n);
Input
    n : non-negative integer;
Output
    $n!$;
Local Variables
    f;
Begin
1   **if** $n = 0$ **then**
2           $f := 1$
3   **else**
4           $f := n * Rec\_fact(n - 1)$;
5   *Return*(f)
End

Figure 3.2. An MPL recursive procedure for $n!$. (Implementation: Maple (txt), Mathematica (txt), MuPAD (txt).)

the operator simulates the looping operation by calling on itself (line 4) to perform a simpler version of the calculation. A procedure that calls itself directly (as in this example) or indirectly through a sequence of procedures is called a *recursive procedure*. The case $n = 0$ (lines 1 and 2) is referred to as a *termination condition* for the procedure since it is defined directly and does not require further calls on *Rec_fact*. As in Equation (3.3), for $n > 0$ the calculation is eventually reduced to the termination condition that stops the process. Each recursive procedure must have one or more termination conditions.

The *Rec_fact* procedure is presented to illustrate simply what is meant by recursion in mathematics and to show how a recursive procedure is expressed in MPL. However, there is more to recursive programming than is shown by this example, and the topic will be discussed in greater detail in Chapter 5.

Recursive Structure of Expressions

One reason recursion is essential to symbolic computation has to do with the *recursive structure* of mathematical expressions. This structure is described using the terms in the following definition:

Definition 3.1. *Mathematical expressions are classified as either atomic expressions or compound expressions:*

1. *An atomic expression is an integer, real, symbol, or reserved symbol (e, ∞, **true**, etc.). The atomic expressions are the atoms or basic building blocks of more involved mathematical expressions.*

2. *A compound expression is composed of an operator with operands. The operator can be an algebraic operator (+, −, etc.), a relational operator (=, <, etc.), a logical operator (**and**, **or**, **not**), a set operator (∪, ∩, ∼), a function or operator name, or the terms **set** or **list**. An operand of an operator can be either an atomic expression or another compound expression. Depending on the operator, each operator can have one or more operands.* □

Example 3.2. Consider the expression $m * x + b$. Since it is common practice in mathematics to give higher precedence to $*$ than $+$, we view the expression as a sum with two operands: the compound expression $m*x$ and the atomic expression b. The operator $*$ has two operands, the atomic expressions m and b. In a similar way, the equation $y = m * x + b$ has the operator $=$ with two operands, the atom y and the compound expression $m * x + b$. □

Example 3.3. The expression $n!$ has one operator $!$ with one operand the symbol n. □

Example 3.4. Consider the expression $Integral(\sin(x), x)$. The $Integral$ operator has two operands, the compound expression $\sin(x)$ and the symbol x. The function $\sin(x)$ has the operator $\sin$ with the single operand the symbol x. □

Example 3.5. Consider the list $[a, b, c]$. In MPL, a list is viewed as a mathematical expression with the term **list** as the operator and the members of the list, a, b, and c, as operands. It may seem odd to think of the term **list** as an operator since it is not as "action oriented" as an operator like $+$. However, this view of lists gives a uniform structure for all compound expressions. In a similar way, the set $\{a, b, c\}$ is a compound expression with operator **set**. □

Definition 3.1 is a recursive description of a mathematical expression since a compound expression is constructed with an operator and simpler expressions (the operands) that are either compound expressions themselves or termination symbols (atomic expressions). Although it may sound as if we are using a recursive definition to state the obvious, we shall see that the recursive structure of expressions implies that a recursive algorithm is appropriate (or even essential) for many mathematical operations.

Recursion in Mathematics

Although the recursion concept is discussed in textbooks on computer science, it is rarely mentioned in textbooks on algebra, trigonometry, and calculus. Therefore, you may find it surprising that you often use recursion when doing pencil and paper manipulations. For example, consider the algebraic simplification of the expression

$$3 \cdot (x + x) + x^2/x \to 7 \cdot x \tag{3.4}$$

using the simplification rules ordinarily found in automatic simplification. To simplify the sum we first simplify each of its operands. And to simplify the first operand

$$3 \cdot (x + x), \tag{3.5}$$

we first simplify each of its operands 3 and $x + x$. The expression 3 is an atom and requires no further simplification. To simplify $x + x$, we first simplify its operands (the two x symbols which are atoms and require no simplification) and then apply a simplification rule to obtain $2 \cdot x$. At this point Expression (3.5) has been transformed to the form $3 \cdot (2 \cdot x)$, and we apply simplification rules to obtain $6 \cdot x$ for the first term in Expression (3.4). In a similar way, we simplify the second term in (3.4) to x and apply the simplification rules to $6 \cdot x + x$ to obtain $7 \cdot x$.

An outline of the *Automatic_simplify* procedure[2] we have used to simplify this expression is shown in Figure 3.3. This simplification process is recursive since a compound expression u is simplified by first applying *Automatic_simplify* (line 4) to each of its operands (the simpler expressions) followed by an application of the appropriate rules. In fact, any mathematical operation that involves a systematic examination of all parts of an expression is most likely recursive.

Recursion can arise in computer algebra for another reason. Many mathematical problems are solved by transforming the original problem into another problem. If the new problem involves the same operation as the original problem, then the process is recursive. For example, consider the evaluation of the indefinite integral $\int x \sin(x^2) \, dx$. Using the substitution $u = x^2$, the integral is transformed to

$$\int x \sin(x^2) \, dx = \int \frac{\sin(u)}{2} \, du.$$

---

[2]The interested reader may consult Cohen [24], Section 3.2, for the full version of the *Automatic_simplify* algorithm.

```
     Procedure  Automatic_simplify(u);
     Input
         u : an algebraic expression;
     Output
         A simplified version of u;
     Begin
1        if  u is an atomic expression then
2            Return(u)
3        else
4            v := the new expression formed by applying the
                   Automatic_simplify procedure to each operand of u;
5            w := the new expression formed by applying the
                   appropriate simplification rules to v;
6            Return(w)
     End
```

Figure 3.3. An outline of an MPL recursive simplification procedure.

To evaluate the original integral, the *Integral* operator must choose the proper substitution and apply itself to a new integral. Since the integration is defined in terms of another (simpler) integration, the process is recursive. In Section 5.3 we describe a recursive algorithm for a basic *Integral* operator that can evaluate integrals similar to the one above.

Exercises

1. Explain why each of the operations can be viewed as a recursive process, and give a termination condition for the recursion:

   (a) The differentiation operation.

   (b) The operation $\lim_{x \to a} f(x)$.

   (c) Polynomial division.

   (d) The expansion of products and powers of polynomials. For example,

   $$\left((x+1)^2 + 2 \right)^2 (x+3) \to x^5 + 7\,x^4 + 22\,x^3 + 42\,x^2 + 45\,x + 27.$$

2. Describe (in words) a recursive algorithm that finds the set of symbols in an expression.

3.2   Expression Structure and Trees[3]

Although Definition 3.1 provides a descriptive language for the recursive structure of an expression, it does not give a scheme for associating a unique structure with an expression. For example, what is the structure of $x + y + z$? Is it the sum of $x + y$ and z, or the sum of x and $y + z$, or even a sum with three operands x, y, and z? Since mathematical expressions are the data objects in computer algebra programming, an understanding of the relationships between their operators and operands is essential. In this section we describe two views of expression structure. The first, which is called the conventional structure, corresponds to the structure in both mathematics and conventional programming languages. The second view, which is called the simplified structure, corresponds to the structure after automatic simplification.

To simplify matters, we focus initially on the algebraic expressions described in the following definition.

Definition 3.6. *An* **algebraic expression** *u is one that satisfies one of the following rules:*

1. *u is an integer.*

2. *u is a symbol.*

3. *u is a sum, product, power, difference, quotient, factorial, or function form, where each operand of u is also an algebraic expression.*   □

The algebraic expressions are the ones we manipulate using the transformation rules of elementary algebra. Notice that the definition is recursive because Rule (3) requires that the operands of a compound algebraic expression are algebraic expressions.

Example 3.7. The expressions

$$2, \quad 1/2, \quad \sin(x), \quad x \wedge 2 + \cos(x), \quad f(x, y, z)$$

are algebraic expressions, while

$$[a, b, c], \quad x + 1 = 2, \quad a \text{ and } b$$

are not.   □

---

[3]In this section, to help clarify expression structure, we use $*$ for the multiplication operator and $\wedge$ for the power operator.

Operator Classification

The following terminology for operators is used to describe expression structure.

Definition 3.8. *Two operators in an algebraic expression u are at* **different parentheses levels** *if one of the operators is inside a pair of matching parentheses, while the other is not. On the other hand, when two operators are not at different levels, they are considered at the* **same parentheses level**.

Example 3.9. In $a*(b+c)$, the operators $*$ and $+$ are at different parentheses levels. In $a * (b + c)/d$, the operators $*$ and $/$ are at the same level, while $+$ is at a different level from either one of them.   □

Operators in an expression are also classified according to the number of operands and the location of the operands relative to an operator. These properties are described with the following terminology:

- A *unary postfix operator* is one with one operand that immediately precedes the operator. For example, in $n!$, the factorial operator is a unary postfix operator.

- A *unary prefix operator* is one with one operand that immediately follows the operator. For example, in $-x$, the difference operator is a unary prefix operator.

- A *function prefix operator* is an expression in function notation with one or more operands. For example, in $f(x, y)$, the function name f is a function prefix operator with two operands x and y.

- A *binary infix operator* is one with two operands, one that immediately precedes the operator and the other that immediately follows the operator. For example, in $a + b$, the $+$ is a binary infix operator. Furthermore, with the conventional view of expressions (described below), both $+$ operators in $a + b + c$ are binary infix operators. In this view, the first $+$ has operands a and b, and the second $+$ has operands $a + b$ and c.

- An *n-ary infix operator* is one with two or more operands that are adjacent to some occurrence of the operator at the same parentheses level. For example, in the simplified view of algebraic expression structure (described below), both $+$ and $*$ are n-ary infix operators. In this view, $a + b + c + d$ is an n-ary sum with four operands a, b, c, and d.

Conventional Structure of Algebraic Expressions

The *conventional structure* of an expression is similar to the structure assumed in both mathematics and conventional programming languages, and, in some computer algebra languages, corresponds to the structure before automatic simplification. The following structural assumptions (Definition 3.10) and precedence rules (Definition 3.11) describe the conventional structure of an expression.

Definition 3.10. (**Structural assumptions for conventional algebraic expressions.**) *Let u be an algebraic expression. The algebraic operators in u satisfy the following structural assumptions:*

1. *The operators + and − are either unary prefix or binary infix operators.*

2. *The operators ∗, /, and ∧ are binary infix operators.*

3. *The operator ! is a unary postfix operator.*

The relationship between operators and operands is defined by the following operator precedence rules.

Definition 3.11. (Conventional precedence rules.) *Let u be an algebraic expression.*

1. *The relative precedence of operators in u at the same parentheses level is given by the precedence hierarchy[4]*

$$
\begin{array}{c}
\textit{(highest level)} \\
\textit{function names} \\
! \\
\wedge \\
*, \, / \\
+, \, - \\
\textit{(lowest level).}
\end{array}
$$

If one operator is below another in the table, that operator has lower precedence. If two algebraic operators in u are at the same level in the table, the relative precedence is determined by the following rules:

(a) *If the operators are + or − operators, ∗ or / operators, or ! operators, then the operator to the right has lower precedence.*

---

[4] Some authors assign unary + and − higher precedence than ∗ and /. We do not make this distinction here.

    (b) If the operators are ∧ operators, then the operator to the left has lower precedence.

2. *For two operators at different parentheses levels, the operator outside a pair of parentheses has lower precedence than the operator inside the parentheses.*

When an expression is adjacent to two operators, it is an operand of the operator with highest precedence.

Example 3.12. In $m * x + b$, the operator $+$ has lower precedence than $*$ which implies the expression is equivalent to $(m * x) + b$.

    In $a * b * c$, both operators operate in a binary fashion and the $*$ on the right has lower precedence than the $*$ on the left. From the conventional viewpoint, this expression is equivalent to $(a * b) * c$.

    In $2 * \sin(x + 1)$, the operator $*$ has lowest precedence, the function name sin is next, and the operator $+$ has highest precedence.

    Finally, in $a * b \wedge c \wedge d * e$, the operators in order of lowest to highest precedence are: the $*$ on the right, the $*$ on the left, the $\wedge$ on the left, and the $\wedge$ on the right. From a conventional viewpoint, this expression is equivalent to $(a * b \wedge (c \wedge d)) * e$.   □

Expression Trees

The structure of an expression comprises the relationships between its operators and operands. An *expression tree* is a diagram that displays this structure. For example, the expression

$$c + d * x \wedge 2 \tag{3.6}$$

is represented by the expression tree in Figure 3.4.

    Each operator and atom in an expression is represented by a position or *node* in the tree. The contents of the nodes and the relationships between the nodes are determined by the operator precedence rules. The operator with lowest precedence in an expression appears at the top of the tree. This top node is called (oddly enough) the *root node* of the tree. According to the precedence rules, the operator $+$ has lowest precedence in $c + d * x \wedge 2$, and so appears at the root. This root operator is also called the *main operator* of the expression, a designation that emphasizes that $c + d * x \wedge 2$ is viewed as a sum with two operands c and $d * x \wedge 2$. The lines that emanate below an operator node connect the operator to each of its operands, and the part of the tree that represents an operand is called a *branch* or *sub-tree*. In this case, the first operand of $+$ is the symbol c, which is represented by the left branch with a node containing c. The right

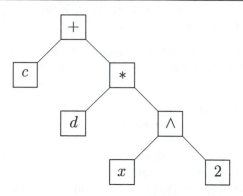

Figure 3.4. The conventional expression tree for $c + d * x \wedge 2$.

branch, which represents the expression $d * x \wedge 2$, is constructed by applying the above process (recursively) to this sub-expression. The main operator of this branch is $*$, and the two branches correspond to the operands d and $x \wedge 2$. Continuing in this fashion, we construct the tree that represents the structure of the expression[5].

Example 3.13. Figure 3.5(a) contains the expression tree for $-a * (b + c)$. Notice that the operator $-$ has lower precedence than $*$ and therefore appears at the root node of the tree.

Figure 3.5(b) contains the expression tree for $Integral(\sin(x), \; x)$. The operator $Integral$, which has lowest precedence, appears at the root node of the tree, and the two operands $\sin(x)$ and x are branches that emanate from this node. In a similar way, the function name sin appears at the root of the branch for $\sin(x)$.

Figure 3.5(c) contains the expression tree for $1/(a * x * y)$. In the conventional view, both $*$ operators act in a binary fashion. □

---

[5] In Mathematica, the operator `TreeForm[u]` displays the tree structure of an expression. However, the displayed structure corresponds to the simplified structure described in Definition 3.14 and Definition 3.16.

In MuPAD, the operator `prog::exprtree` displays the tree structure. The conventional structure is obtained with `prog::exprtree(hold(u))`, while the simplified structure (described in Definition 3.14 and Definition 3.16) is obtained with `prog::exprtree(u)`. In MuPAD, the conventional structure is similar to the structure described here with two modifications. First, both $+$ operators and $*$ operators are represented as n-ary operators. Next, numerical fractions have a special form with the operator DOM_RAT at the root.

The current release of Maple (7) does not have an operator to display the tree. However, the tree structure can be obtained using the structure operators (see Figure 3.18 on page 106).

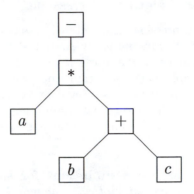

(a) A conventional expression tree for $-a * (b + c)$.

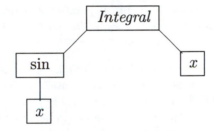

(b) A conventional expression tree for $Integral(\sin(x),\ x)$.

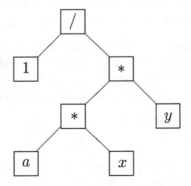

(c) A conventional expression tree for $1/(a * x * y)$.

Figure 3.5. Conventional expression trees.

Simplified Structure of Algebraic Expressions

Since mathematical expressions encountered as data objects in computer algebra programs are in automatically simplified form, the structure of these expressions is particularly important to us. This simplified structure (described below) simplifies the programming process by eliminating extraneous operators from an expression and by providing easier access to its operands. The following structural assumptions (Definition 3.14) and precedence rules (Definition 3.16) describe some properties of simplified expressions.

Definition 3.14. (**Structural assumptions for simplified algebraic expressions**) *Let u be an automatically simplified algebraic expression. The operators in u satisfy the following structural assumptions:*

> *1. The operator $+$ is an n-ary infix operator with two or more operands and none of its operands is a sum. In addition, at most one operand of $+$ is an integer or fraction.*

In the simplified view, the expression $a+b+c$ is viewed as a sum with three operands a, b, and c rather than the conventional view as a binary sum with operands $a+b$ and c (see Figure 3.6(a)). Furthermore, the expression $a+(b+c)$ is not in automatically simplified form because one of the operands of the main operator $+$ is also a sum. Indeed, the automatically simplified form of this expression is $a+b+c$. Finally, Rule (1) implies that the unary sum $+x$ is not automatically simplified because a sum must have at least two operands. The simplified form of $+x$ is x.

> *2. The operator $*$ is an n-ary infix operator with two or more operands and none of its operands is a product. In addition, at most one operand of $*$ is an integer or fraction, and when an integer or fraction is an operand of a product, it is the first operand [6].*

This rule implies the simplified form of $a*2*(b*c)$ is the n-ary product $2*a*b*c$.

> *3. The unary operator $-$ and the binary operator $-$ do not appear in simplified expressions.*

---

[6] In both Maple and Mathematica, an integer or fraction operand in a product is the first operand.

In MuPAD, however, an integer or fraction operand in a product is represented internally as the last operand even though the displayed form indicates it is the first operand. Since some algorithms in later chapters assume an integer or fraction in a product is the first operand, the MuPAD implementations are modified to account for this difference.

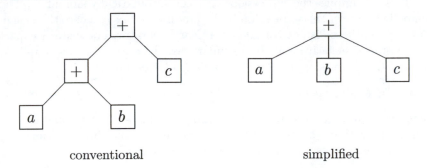

conventional   simplified

(a) Conventional and simplified structures for $a + b + c$.

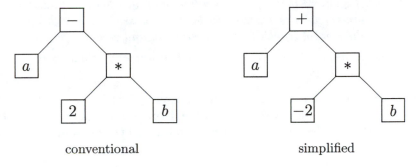

conventional   simplified

(b) Expression trees for $a - 2 * b$ and its simplified form $a + (-2) * b$.

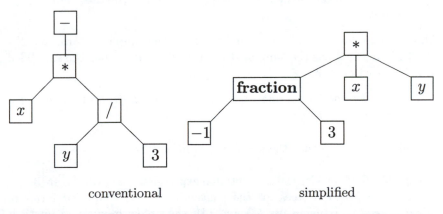

conventional   simplified

(c) Expression trees for $-x * y/3$ and its simplified form $(-1/3) * x * y$.

Figure 3.6. Conventional and simplified expression trees.

This rule implies the expression $-x$ is not automatically simplified. The simplified form of this expression is the product $(-1) * x$, where the $-$ sign in the parentheses is not considered a unary operator, but instead is part of the integer negative one. In a similar way, the automatically simplified form of the expression $a - 2 * b$ is $a + (-2) * b$ (see Figure 3.6(b)).

4. The binary operator / does not appear in simplified expressions.

Simplified quotients are represented using products, powers, and special forms for fractions (described below). For example, the simplified form for $1/c$ is c^{-1} and the simplified form for c/d^2 is $c * d^{-2}$.

5. Numerical fractions satisfy the following:

   (a) *A quotient that represents a fraction c/d, where $c \neq 0$ and $d \neq 0$ are integers is represented by an expression tree with operator the symbol* **fraction**, *first operand c, and second operand d.*

   (b) *A negative fraction has a negative numerator and positive denominator.*

This rule implies that the expression $1/2$ is not viewed as a quotient but instead as an expression with operator **fraction** and operands 1 and 2. In addition, the expression $-1/2$ has the operator **fraction** and operands -1 and 2.

   6. The operator $\wedge$ is a binary operator. In addition, if $u = v \wedge n$, where n is an integer, then v cannot be an integer, fraction, product, or power.

This rules implies the following: $(a * b) \wedge 2$ has the simplified form $(a \wedge 2) * (b \wedge 2)$; $(x \wedge 2) \wedge 3$ has the simplified form $x \wedge 6$; and 3^{-1} has the simplified form the fraction $1/3$.

   7. The operator ! is a unary postfix operator whose operand is not a non-negative integer.

This rule implies that the simplified form of 3! is 6.

Example 3.15. Consider the conventional expression $-x * y/3$. Figure 3.6(c) shows the expression trees for the conventional and simplified forms of this expression. Using Rules (3) and (4), the unary minus and quotient operators are removed. Therefore, using the Rules (2), (5), and (6), the simplified structure is $((-1)/3) * x * y$, where $*$ is an n-ary operator with three operands, the $-$ is part of the integer -1, and the fraction $(-1)/3$ has main operator **fraction**. $\square$

Definition 3.16. (**Simplified precedence rules**) *Let u be an automatically simplified algebraic expression.*

1. *The relative precedence of operators in u at the same parentheses level is given by the precedence hierarchy*

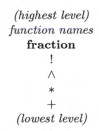

   (highest level)
   function names
   fraction
   !
   ∧
   *
   +
   (lowest level)

   If two ! operators are at the same parentheses level, then the one to the right has lower precedence. If two ∧ operators are at the same parentheses level, then one to the left has lower precedence.

2. *For operators at different parentheses levels, operators outside a pair of parentheses have lower precedence than operators inside the parentheses.*

Because of the structural assumptions, the precedence rules for simplified expressions are simpler than those for conventional expressions. For example, since multiple occurrences of the + operator at the same parentheses level coalesce to a single operator in an expression tree, there is no need to account for this situation in the precedence rules. Notice that the **fraction** operator has higher precedence than *, ∧, and ! so that a fraction is isolated as an operand relative to these operators. This point is illustrated in Figure 3.6(c).

Using the structural assumptions in Definition 3.14 and the precedence rules in Definition 3.16, the definition of an algebraic expression can be modified in the following way.

Definition 3.17. *An expression u is an* **automatically simplified algebraic expression (ASAE)** *if it satisfies one of the following rules:*

1. *u is an integer.*

2. *u is a fraction c/d where c ≠ 0, d ≠ 0 are integers.*

3. *u is a symbol.*

4. u is a sum, product, power, factorial, or function form, where each operand of u is also an automatically simplified algebraic expression.

Although the structural assumptions and precedence rules describe some important properties of automatically simplified expressions, they are not by any means a complete description of these expressions. For example, the structural assumptions do not describe all the ordering properties of operands in sums and products, the properties of powers or the special rules involving the integers 0 and 1. Since a complete description of automatically simplified expressions is quite involved, it is not included here[7]. At this point our intent is to give a description that is sufficient to begin computer algebra programming. Additional structural rules can be found by experimentation with computer algebra software (see Exercise 3).

Most computer algebra systems use an internal form for automatically simplified algebraic expressions that is similar to the one described here, although the displayed form may disguise the actual structure. We illustrate this point in Figure 3.7, which shows a Mathematica session together with the simplified structure of $a/b + c - d$. The command `TreeForm` at $In[2]$ displays a representation of the expression tree. Observe that the displayed form of the simplified expression at $Out[1]$ includes the quotient and difference operators, even though the Mathematica internal tree structure does not.

Since the simplified structure of an expression may not be apparent from its displayed form, an operator may transform an expression in an unexpected way. To illustrate this point, consider the Maple dialogue in Figure 3.8 that obtains the derivative of $f(x)/g(x)$. Since the simplified form of this expression is $f(x) * g(x)^{(-1)}$, the derivative is obtained with the differentiation product and power rules rather than the quotient rule. Similar results are obtained with both Mathematica and MuPAD.

Functions Transformations for Exponential Functions and Powers in Automatic Simplification

In ordinary mathematical notation, the expressions e^x and $\exp(x)$ are two forms that are used to represent the exponential function. Some computer algebra systems allow both representations for input but may use only one form in automatically simplified expressions. In a similar way, a CAS may allow both $\sqrt{x}$ (or sqrt(x)) and $x^{1/2}$ for the square root function, but

---

[7]For this description, consult Cohen [24], Chapter 3, where a complete description is needed for an algorithm that transforms an algebraic expression to automatically simplified form.

$In[1] :=$ a/b + c − d

$Out[1] =$ $\frac{a}{b}$ + c − d

$In[2] :=$ TreeForm[a/b + c − d]

$Out[2]$ //TreeForm =
     Plus [| , c, |]
         Times [a, |]   Times [−1, d]
             Power [b, −1]

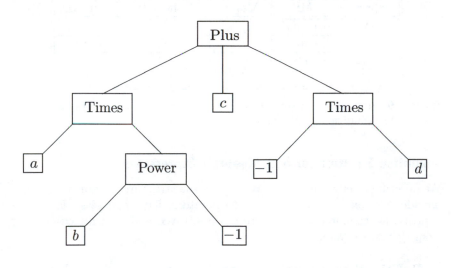

Figure 3.7. A Mathematica session with the simplified structure for the expression $a/b + c − d$.

choose one internal simplified form for both. In MPL, we represent the exponential function with the function form $\exp(x)$ and the square root function with the power $x^{1/2}$. In Figure 3.9, we give the representations of these functions in Maple, Mathematica, and MuPAD.

```
> u := f(x)/g(x);
```

$$u := \frac{f(x)}{g(x)}$$

```
> diff(u, x);
```

$$\frac{\frac{\partial}{\partial x} f(x)}{g(x)} - \frac{f(x)\left(\frac{\partial}{\partial x} g(x)\right)}{g(x)^2}$$

Figure 3.8. A Maple dialogue that shows that the derivative of $f(x)/g(x)$ is obtained with the product rule and power rule.

expression	MPL	Maple	Mathematica	MuPAD
$\exp(x)$	$\exp(x)$	$\texttt{exp(x)}$	E^x	$\texttt{exp(x)}$
e^x	$\exp(x)$	$\texttt{exp(1)}^x$	E^x	$\texttt{exp(x)}$
$\sqrt{x}$	$x^{1/2}$	$x^{1/2}$	$x^{1/2}$	$x^{1/2}$
$x^{1/2}$	$x^{1/2}$	$x^{1/2}$	$x^{1/2}$	$x^{1/2}$

Figure 3.9. Simplified structure of the exponential and power functions in MPL, Maple, Mathematica, and MuPAD.

Simplified Structure for Non-Algebraic Expressions

We describe briefly some issues related to the structure of expressions that include relational operators, logical operators, lists, and sets. For these expressions, the internal forms are more involved and, in some cases, vary from system to system.

Relational expressions. For expressions with one relational operator ($=$, $<$, etc.), most systems use a binary structure with relative precedence levels for relational operators below the levels for algebraic operators. Figure 3.10 gives the simplified structure for the expression for $x + 1 < 2 * x$. Maple, Mathematica, and MuPAD use a similar internal form for this expression.

Both the Mathematica and MuPAD systems provide for more involved relational expressions that contain two or more relational operators. For example, in Figure 3.11 we show the internal form for the expression $x < y < z$ in these systems. Observe that Mathematica represents the expression using an n-ary form, while MuPAD uses a nested binary form.

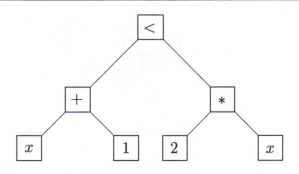

Figure 3.10. The MPL simplified structure for the expression $x + 1 < 2 * x$.

Figure 3.12 gives the internal forms in Mathematica and MuPAD for the expression $x \leq y < z$, which has two different relational operators. Observe that Mathematica represents the expression using a form that includes relational operators as operands of the main operator `Inequality`, while MuPAD represents this expression using a nested binary form.

Finally, in both the Maple and MuPAD systems, expressions with $>$ and $\geq$ are converted by automatic simplification to equivalent expressions with $<$ and $\leq$. For example, in both of these systems, $a > b$ is simplified to $b < a$.

Logical expressions. The relative precedence levels of the logical operators is

(lowest)

not

and

or

(highest),

and the logical operators have lower precedence than relational operators.

Although logical expressions are used primarily as Boolean tests in both decision and looping structures, some computer algebra languages allow logical expressions as program statements or data objects. For example, the expression

p **or not** q **and** r

is a mathematical expression with structure shown in Figure 3.13(a).

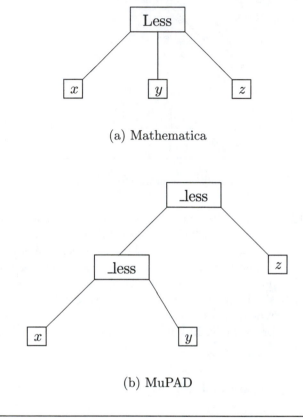

(a) Mathematica

(b) MuPAD

Figure 3.11. The internal form for $x < y < z$ in Mathematica and MuPAD. (Implementation: Mathematica (nb), MuPAD (mnb).)

In Figure 3.13(b), we show the structure of an expression with logical, relational, and algebraic operators. Maple, Mathematica, and MuPAD allow expressions to be used in this way[8].

In Mathematica and MuPAD, logical expressions have simplified forms in which both the **and** and **or** operators are n-ary infix operators with

---

[8] In Maple, there is one curious exception to this statement. Suppose x, y, and z are unassigned symbols. Although, in this system, the expression x < y and y < z is not changed by automatic simplification, the similar expression x < y and y = z is transformed to **false**. This occurs since, in this context, y = z evaluates to **false** because y and z are distinct symbols. On the other hand, the expression y = z by itself remains unchanged by automatic simplification.

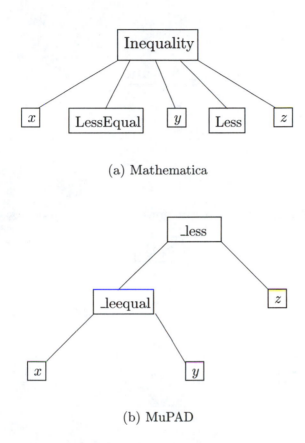

(a) Mathematica

(b) MuPAD

Figure 3.12. The internal form for the expression $x \leq y < z$ in Mathematica and MuPAD. (Implementation: Mathematica (nb), MuPAD (mnb).)

operands of a different type. (For example, the operator **and** cannot have an operand that is also an **and**.) In Figure 3.14(a),(b), we give Mathematica and MuPAD representations of the simplified form of the expression

$$w \text{ or } x \text{ and } y \text{ or not } z \qquad (3.7)$$

which has two **or** operators at the same level.

On the other hand, in the Maple system, the simplified forms for **and** and **or** remain as binary operators. The internal form for Expression (3.7) in this system is shown in Figure 3.14(c).

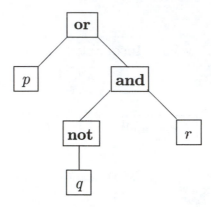

(a) The MPL simplified structure for the expression p **or not** q **and** r.

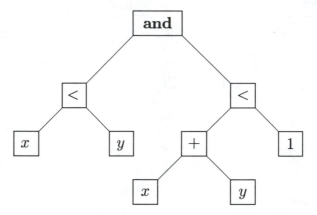

(b) The MPL simplified structure for the expression $x < y$ **and** $x + y < 1$.

Figure 3.13. MPL structures for logical expressions.

Sets and Lists. In MPL, both sets and lists are viewed as expressions with main operator **set** or **list** along with operands that are the expressions in the set or the list. In addition, the expression tree for the empty set $\emptyset$ or empty list [ ] is the tree with a single node **set** or **list**. Figures 3.15(a),(b) illustrate the tree structures for expressions with sets and lists. Both Maple and MuPAD represent sets and lists in this way. Mathematica, which uses the brace notation { and } for both sets and lists, uses the symbol List as the main operator for these expressions

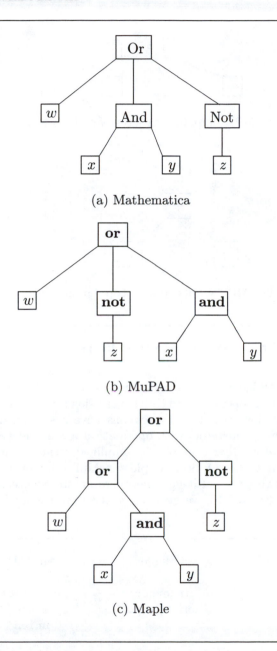

(a) Mathematica

(b) MuPAD

(c) Maple

Figure 3.14. The simplified structure of the expression w **or** x **and** y **or** **not** z in Mathematica, MuPAD, and Maple.

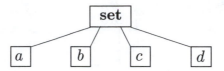

(a) The MPL simplified structure for the set $\{a, b, c, d\}$.

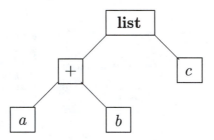

(b) The MPL simplified structure for the list $[a + b, c]$.

Figure 3.15. Expression tress for a set and a list.

In some CAS languages, the set operators ($\cup$, $\cap$, $\sim$), can have symbols (instead of sets) as operands and these expressions can act as data objects. When used in this way, these expressions have a simplified form where $\cup$ and $\cap$ are n-ary operators with operands of a different type, and $\sim$ is a binary operator. (For example, in simplified form, $\cup$ cannot have an operand that is also a $\cup$.) Both Maple and MuPAD allow set expressions to be used in this way, although the operator precedence levels are not the same. In Figure 3.16, we give the relative precedence levels of these

MPL	Maple	MuPAD
$\cap$	intersect	intersect
$\sim$	union, minus	minus
$\cup$		union

Figure 3.16. The relative precedence of set operations in MPL, Maple, and MuPAD.

operators in MPL, Maple, and MuPAD. Observe that in both MPL and MuPAD the operators are at three different levels, while in Maple union and minus are at the same level.

In Figure 3.17, we give the simplified representations in Maple and MuPAD for the expression

$$A \cup B \cup C \cap D \sim E.$$

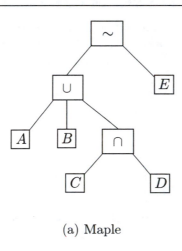

(a) Maple

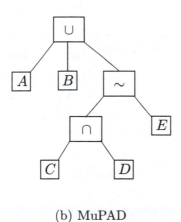

(b) MuPAD

Figure 3.17. The structure of the expression $A \cup B \cup C \cap D \sim E$ in Maple and MuPAD.

Observe that the meaning of the expression in Maple is

$$(A \cup B \cup (C \cap D)) \sim E,$$

while the meaning in MuPAD is

$$A \cup B \cup ((C \cap D) \sim E).$$

Primitive Operations on Simplified Mathematical Expressions

In order to analyze and manipulate a mathematical expression, we must access its operators and operands. MPL uses three primitive operators to perform these tasks.

Definition 3.18. *The operator*

$$Kind(u)$$

is defined by the following rules:

1. *If u is an atomic expression, $Kind(u)$ returns the type of expression (e.g., **integer**, **real**, or **symbol**).*

2. *If u is a compound expression, $Kind(u)$ returns the operator at the root of the expression tree.*

Example 3.19.

$$
\begin{aligned}
Kind(x) &\rightarrow \textbf{symbol}, \\
Kind(3) &\rightarrow \textbf{integer}, \\
Kind(2.1) &\rightarrow \textbf{real}, \\
Kind(\pi) &\rightarrow \textbf{symbol}, \\
Kind(m * x + b) &\rightarrow \textbf{+}, \\
Kind((a + b) * \sin(x \wedge 2)) &\rightarrow \textbf{*}, \\
Kind((a/b)) &\rightarrow \textbf{*}, \\
Kind(2/3) &\rightarrow \textbf{fraction}, \\
Kind(\sin(x)) &\rightarrow \textbf{sin}, \\
Kind(a = b) &\rightarrow \textbf{=}, \\
Kind(\{a, b, c, d\}) &\rightarrow \textbf{set}, \\
Kind(x \textbf{ and } y) &\rightarrow \textbf{and}, \\
Kind(x - x + 2) &\rightarrow \textbf{integer}.
\end{aligned}
$$

In the last example, the operand is simplified by automatic simplification to the integer 2. (Implementation: Maple (mws), Mathematica (nb), MuPAD (mnb).) □

Definition 3.20. *If u is a compound expression, the operator*

$$Number\_of\_operands(u)$$

returns the number of operands of the main operator of u. If u is not a compound expression, then Number_of_operands returns the global symbol **Undefined***.*

Example 3.21.

$$
\begin{aligned}
Number\_of\_operands(m*x+b) &\rightarrow 2, \\
Number\_of\_operands(f(x,y)) &\rightarrow 2, \\
Number\_of\_operands(\{a,b,c,d\}) &\rightarrow 4, \\
Number\_of\_operands(n!) &\rightarrow 1, \\
Number\_of\_operands(x) &\rightarrow \textbf{Undefined}.
\end{aligned}
$$

In the last example, the input expression x is not a compound expression. (Implementation: Maple (mws), Mathematica (nb), MuPAD (mnb).) □

Definition 3.22. *If u is a compound expression, the operator*

$$Operand(u,i)$$

returns the ith operand of u. If u is not a compound expression or u does not have an ith operand, then Operand returns the global symbol **Undefined***.*

Example 3.23.

$$
\begin{aligned}
Operand(m*x+b,2) &\rightarrow b, \\
Operand(x \wedge 2,1) &\rightarrow x, \\
Operand(Operand(m*x+b,1),2) &\rightarrow x, \\
Operand(\{a,b,c,d\},2) &\rightarrow b, \\
Operand(x-x,1) &\rightarrow \textbf{Undefined}, \\
Operand(2/(-3),2) &\rightarrow 3.
\end{aligned}
$$

The last two examples are based on the simplified form of the expression. (Implementation: Maple (mws), Mathematica (nb), MuPAD (mnb).)

MPL	Maple	Mathematica	MuPAD
$Kind(u)$	whattype(u) and op(0,u) for function names	Head(u)	type(u) and op(u,0) for undefined function names
$Operand(u,i)$	op(i,u)	Part[u,i] and Numerator[u] and Denominator[u] for fractions	op(u,i)
$Number\_of\_operands(u)$	nops(u)	Length[u]	nops(u)
$Construct(f,L)$	see Figure 3.19	Apply[f,L]	see Figure 3.20

Figure 3.18. The primitive MPL structural operators in Maple, Mathematica, and MuPAD.

Keep in mind, because automatic simplification in a computer algebra system may apply the commutative law to reorder the operands in a sum or product, the *Operand* operator may obtain an unexpected result. For example, if $b + a$ is reordered to $a + b$, we obtain

$$Operand(b + a,\ 2) \to b. \qquad \Box$$

The operators *Kind*, *Number_of_operands*, and *Operand* are the three basic operations that are used to analyze and manipulate mathematical expressions, and most computer algebra systems have versions of these operators (see Figure 3.18).

Construction of Expressions

In some instances, we need to construct an expression with a given operator and list of operands. The MPL operator *Construct* is used for this purpose.

Definition 3.24. *Let f be an operator ($+$, $*$, $=$, etc.) or a symbol, and let $L = [a, b, \ldots, c]$ be a list of expressions. The operator*

$$Construct(f, L)$$

returns an expression with main operator f and operands $a, b, \ldots, c$.

Example 3.25.

$$Construct(" + ", [a, b, c]) \quad \to \quad a + b + c,$$
$$Construct(" * ", [a + b, c + d, e + f]) \quad \to \quad (a + b) * (c + d) * (e + f),$$
$$Construct(g, [a, b, c]) \quad \to \quad g(a, b, c),$$

(Implementation: Maple (mws), Mathematica (nb), MuPAD (mnb).) $\Box$

```
Construct := proc(f,L)
local g,s;
  if f = '!' then RETURN(op(L)!);
  elif member(f,{'and','or'}) then RETURN(convert(L,f))
  elif f = 'not' then RETURN(not op(L))
  elif f = set then RETURN({op(L)})
  elif f = list then RETURN(L)
  else s := subsop(0=f,g(op(L))); RETURN(eval(s))
  fi
end:
```

Figure 3.19. A Maple procedure to implement MPL's *Construct* operator. (Implementation: Maple (txt).)

```
Construct := proc(f,L)
begin
  if f = _divide then return(op(L,1)/op(L,2))
  elif f = _subtract of f = _negate then return(op(L,1)-op(L,2))
  elif f = DOM_SET then return({op(L)})
  elif f = DOM_LIST then return(L)
  else return(f(op(L)))
  end_if
end_proc:
```

Figure 3.20. A MuPAD procedure to implement MPL's *Construct* operator. (Implementation: MuPAD (txt).)

While Mathematica has an operator that constructs expressions (see Figure 3.18), Maple and MuPAD do not. However, in both of these languages, the operation can be simulated with a procedure (see Figures 3.19 and 3.20).

Exercises

1. Experiment with a CAS to determine the simplified structure of the expressions in Figure 3.6. Are the structures the same as those shown in the figure?

2. For each of the following, give the conventional and simplified structures of the expression. In addition, for each of the expressions, compare the simplified structure based on the rules in the text to the simplified internal structure in a CAS.

   (a) $a/b - c/d$.

   (b) $(x \wedge a) \wedge 3$.

   (c) $(x \wedge 2) \wedge (1/2)$.

   (d) $\dfrac{3}{2 * a * (x - 1)}$.

   (e) $(-2 * x) \wedge (-3)$.

   (f) $((x - y) + z) + w$.

   (g) $((x - y) * y/2) \wedge 2$.

   (h) $\dfrac{x \wedge 2 - 1}{x - 1}$.

   (i) $\dfrac{1}{x \wedge y}$.

   (j) $\dfrac{-x * (a/b)}{c}$.

   (k) $\dfrac{2}{a + b} * \dfrac{3}{c + d}$.

   (l) $x \wedge 2 + \cos(1/x - 2)$.

   (m) $x = \dfrac{-b + (b \wedge 2 - 4 * a * c) \wedge (1/2)}{2 * a}$.

3. Experiment with a CAS to determine some additional structural rules that describe automatically simplified algebraic expressions. For example:

   (a) Can a sum or product have two identical operands?

   (b) How are operands in sums and products combined?

   (c) Can the first operand of the power operator $\wedge$ also be a power?

   (d) What are the special rules that involve 0 and 1? For example, can a 0 or 1 be an operand of a sum, product, power, or factorial?

   Exercise 1 on page 55 is helpful for this exercise. For further discussion of structural assumptions for automatically simplified algebraic expressions, see Cohen [24], Chapter 3.

3.3 Structure-Based Operators

In this section we describe four operators for which the operations are based only on the simplified structure of an expression. First, we introduce the terminology that is used in the definitions of these operators.

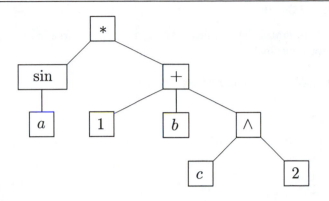

Figure 3.21. A simplified expression tree for $\sin(a) * (1 + b + c \wedge 2)$.

Complete Sub-Expressions

Definition 3.26. *Let u be an automatically simplified expression. A **complete sub-expression** of u is either the expression u itself or an operand of some operator in u.*

In terms of expression trees, the complete sub-expressions of u are either the expression tree for u or one of its sub-trees.

Example 3.27. Consider the expression

$$\sin(a) * (1 + b + c \wedge 2), \tag{3.8}$$

which has the expression tree shown Figure 3.21. This expression contains the following complete sub-expressions:

$$\sin(a) * (1 + b + c \wedge 2), \quad \sin(a), \quad a,$$
$$1 + b + c \wedge 2, \quad 1, \quad b, \quad c \wedge 2, \quad c, \quad 2.$$

There are some parts of an expression that are sub-expressions in a mathematical sense but are not complete sub-expressions. For example in Expression (3.8), $1 + b$ is not a complete sub-expression since it is not the operand of an operator. $\qquad\square$

The *Free_of* Operator

The *Free_of* operator determines if an expression u is free of an expression t (or does not contain t).

Definition 3.28. *Let u and t (for target) be mathematical expressions. The operator*

$$Free\_of(u, t)$$

returns **false** *when t is identical to some complete sub-expression of u and otherwise returns* **true**.

Example 3.29.

$$
\begin{aligned}
Free\_of(a + b,\ b) &\rightarrow \textbf{false}, \\
Free\_of(a + b,\ c) &\rightarrow \textbf{true}, \\
Free\_of((a + b) * c,\ a + b) &\rightarrow \textbf{false}, \\
Free\_of(\sin(x) + 2 * x,\ \sin(x)) &\rightarrow \textbf{false}, \\
Free\_of((a + b + c) * d,\ a + b) &\rightarrow \textbf{true}, \quad (3.9) \\
Free\_of((y + 2 * x - y)/x,\ x) &\rightarrow \textbf{true}, \quad (3.10) \\
Free\_of((x * y)^2,\ x * y) &\rightarrow \textbf{true}. \quad (3.11)
\end{aligned}
$$

In Statement (3.9), $a+b$ is not a complete sub-expression of $(a+b+c)*d$ and so the operator returns **true**. In Statement (3.10), automatic simplification simplifies the first operand to 2 and so the expression no longer contains an x. In a similar way, in Statement (3.11) automatic simplification transforms $(x * y)^2$ to $x^2 * y^2$ which gives the output **true**. (Implementation: Maple (mws), Mathematica (nb), MuPAD (mnb).)   □

To perform *Free_of*(u, t), each complete sub-expression of u is checked to determine if it is structurally identical to the target t. This is easily done using a recursive search through the expression tree. Briefly, the process goes as follows: first compare u to t, and if $u = t$ the search is done and **false** is returned. If $u \neq t$ and u is an atom, there is nowhere else to search and so **true** is returned. On the other hand, if $u \neq t$ and u is a compound expression, the search continues by recursively applying the process just described to each of the (simpler) operands of u. Continuing in this fashion, we compare t to each of the complete sub-expressions of u. For example, if $u = 3 * x + y * (z + 2)$, the scheme compares a target t to the complete sub-expressions of u in the following order:

$$3 * x + y * (z + 2),\ 3 * x,\ 3,\ x,\ y * (z + 2),\ y,\ z + 2,\ z,\ 2.$$

MPL	Maple	Mathematica	MuPAD
$Free\_of(u, t)$	`not(has(u,t))`	`FreeQ[u,x]`	`not(has(u,t))`
$Substitute(u, t = r)$	`subs(t=r,u)`	`ReplaceAll[u,t->r]` or `u/.t->r`	`subs(u,t=r)`
$Sequential\_substitute$ $(u, [t_1 = r_1, t_2 = r_2])$	`subs(`$t_1 = r_1,$ $t_2 = r_2$`, u)`	`u/.`$t_1 ->$ `r`$_1$ `/.`$t_2 ->$ `r`$_2$	`subs(u,`$t_1 = r_1,$ $t_2 = r_2$`)`
$Concurrent\_substitute$ $(u, \{t_1=r_1, t_2 = r_2\})$	`subs([`$t_1 = r_1,$ $t_2 = r_2$`], u)`	`u/. {`$t_1 -> r_1,$ $t_2 -> r_2$`}`	`subs(u, [`$t_1 = r_1,$ $t_2 = r_2$`])`

Figure 3.22. Structural operators in Maple, Mathematica, and MuPAD that correspond most closely to MPL's structural operators.

An MPL procedure for the *Free_of* operator is given in Section 5.2. An operator similar to *Free_of* is available in most computer algebra systems (see Figure 3.22).

The *Substitute* Operator

Substitution is one of the essential operations used to manipulate and simplify mathematical expressions. The *Substitute* operator performs a particularly simple form of substitution, called *structural substitution*, that is based solely on the tree structure of an expression.

Definition 3.30. *Let u, t, and r be mathematical expressions. The structural substitution operator has the form*

$$Substitute(u, \ t = r).$$

It forms a new expression with each occurrence of the target expression t in u replaced by the replacement expression r. The substitution occurs whenever t is structurally identical to a complete sub-expression of u.

Keep in mind that *Substitute* does not change u, but instead creates an entirely new expression. Some examples of the use of the operator are given in the MPL dialogue in Figure 3.23.

The statements at <1>, <2>, and <3> illustrate that u is not changed by the substitution operation. In <6>, the substitution does not occur since $a + b$ is not a complete sub-expression of $a + b + c$. However, in <7>, we obtain the substitution intended in <6> by modifying the form of the substitution.

Like the *Free_of* operator, the *Substitute* operator searches an expression u in a recursive manner and compares each complete sub-expression

---

<1> $u := a + b;$

$\rightarrow \quad u := a + b$

<2> $v := Substitute(u, \ b = x);$

$\rightarrow \quad v := a + x$

<3> $u/v;$

$\rightarrow \quad \dfrac{a+b}{a+x}$

<4> $Substitute(1/a + a, \ a = x);$

$\rightarrow \quad \dfrac{1}{x} + x$

<5> $Substitute((a+b)^2 + 1, \ a + b = x);$

$\rightarrow \quad x^2 + 1$

<6> $Substitute(a + b + c, \ a + b = x);$

$\rightarrow \quad a + b + c$

<7> $Substitute(a + b + c, \ a = x - b);$

$\rightarrow \quad x + c$

Figure 3.23. An MPL dialogue that illustrates the use of the *Substitute* operator. (Implementation: Maple (mws), Mathematica (nb), MuPAD (mnb).)

of u to the target t. (See Exercise 10 on page 195 for an MPL procedure for the operator.) Most computer algebra systems have a form of the *Substitute* operator (see Figure 3.22).

Substitution and Evaluation

In both the Maple and MuPAD systems, structural substitution may return an expression with an operator in unevaluated form. This point is illustrated in the Maple dialogue in Figure 3.24, which gives a sequence of statements to verify the solution of a differential equation. At the first prompt, we assign a differential equation to w, and at the second prompt, substitute a specific function for $y(x)$. Notice that the differentiation operator in the second display has not been evaluated. At the third prompt, we force this evaluation by applying Maple's `eval` operator to the output of the `subs` operator. The MuPAD system also requires this forced evaluation.

```
> w := diff(y(x),x) + 2*y(x)=5*sin(x);
```

$$w := \left(\frac{\partial}{\partial x} \, \mathrm{y}(x) \right) + 2 \, \mathrm{y}(x) = 5 \sin(x)$$

```
> subs(y(x) = -cos(x)+2*sin(x)+exp(-2*x),w);
```

$$\left(\frac{\partial}{\partial x} \left(-\cos(x) + 2 \sin(x) + e^{(-2\,x)} \right) \right) - 2 \cos(x) + 4 \sin(x) + 2 \, e^{(-2\,x)} = 5 \sin(x)$$

```
> eval(subs(y(x) = -cos(x)+2*sin(x)+exp(-2*x),w));
```

$$5 \sin(x) = 5 \sin(x)$$

Figure 3.24. A Maple dialogue that verifies the solution of a differential equation. (Implementation: Maple (mws), MuPAD (mnb).)

Although the Mathematica system does not, in general, require a forced evaluation after substitution, the substitution of a specific function for an undefined function in a derivative requires a special form. This point is illustrated in the Mathematica dialogue in Figure 3.25. At *In[1]*, we assign a differential equation to w, and at *In[2]*, use the ReplaceAll operator to substitute a specific function in the differential equation. Notice that the substitution has not occurred in the derivative term y'[x]. The issue here is that Mathematica represents y[x] with the internal form Derivative[1][y][x] which does not contain the function form y[x]. To obtain the substitution, it is necessary to represent the specific function in a form that Mathematica calls a *pure function*:

$$\mathrm{Function}[x, \, -\mathrm{Cos}[x] + 2 * \mathrm{Sin}[x] + \mathrm{Exp}[-2 * x]],$$

and substitute this expression for the function name y. We have performed this substitution at *In[3]*, and then at *In[4]*, verified the solution of the differential equation by using the Expand operator to simplify the left side of the previous equation[9].

---

[9] The Expand operator in *In[4]* is not required in either Maple or MuPAD. In both of these systems, automatic simplification obtains the distributive transformation

$$2 \, (\exp(-2\,x) - \cos(x) + 2 \, \sin(x)) \to 2 \, \exp(-2\,x) - 2 \, \cos(x) + 4 \, \sin(x).$$

This transformation is not obtained by automatic simplification in Mathematica.

$In[1]:=$ w $=$ D[y[x], x] $+ 2 *$ y[x] $== 5 *$ Sin[x]

$Out[1]=$ $2\,y[x] + y'[x] == 5\,Sin[x]$

$In[2]:=$ ReplaceAll[w, y[x] $\rightarrow$ $-$Cos[x] $+ 2 *$ Sin[x] $+$ Exp[$-2 *$ x]]

$Out[2]=$ $2\,(e^{-2x} - Cos[x] + 2 * Sin[x]) + y'[x] == 5\,Sin[x]$

$In[3]:=$ z $=$ ReplaceAll[w, y $\rightarrow$ Function[x, $-$Cos[x] $+ 2 *$ Sin[x] $+$ Exp[$-2 *$ x]]]

$Out[3]=$ $-2\,e^{-2x} + 2\,Cos[x] + Sin[x] + 2\,(e^{-2x} - Cos[x] + 2 * Sin[x]) == 5\,Sin[x]$

$In[4]:=$ Expand[Part[z, 1]]

$Out[4]=$ $5\,Sin[x]$

Figure 3.25. A Mathematica dialogue that verifies the solution of a differential equation. (Implementation: Mathematica (nb).)

Multiple Substitution

A *multiple structural substitution* is one in which a collection of structural substitutions is applied to an expression with a single operation. Since the individual substitutions may not be independent (i.e., one substitution may affect the action of another one), both the order of the substitutions and the mechanics of the process may affect the result. We describe below two models for multiple substitution, *sequential substitution* and *concurrent substitution*.

Definition 3.31. *Let u be an expression and let L be a list of equations*

$$L = [t_1 = r_1, t_2 = r_2, \ldots, t_n = r_n]$$

*where the targets t_i are distinct. The **sequential structural substitution** operator has the form*

$$Sequential\_substitute(u, L).$$

The operator returns the expression u_n that is defined by the sequence of structural substitutions

$$u_1 \quad := \quad Substitute(u, \ t_1 = r_1);$$

$$u_2 \quad := \quad Substitute(u_1, \ t_2 = r_2);$$

$$\vdots$$

$$u_n \quad := \quad Substitute(u_{n-1}, \ t_n = r_n);$$

Example 3.32.

$Sequential\_substitute(x + y, \ [x = a + 1, \ y = b + 2]) \rightarrow a + b + 3,$

$Sequential\_substitute(x + y, \ [x = a + 1, \ a = b + 2]) \rightarrow b + 3 + y,$

$Sequential\_substitute(f(x) = a * x + b, \ [f(x) = 2, \ x = 3])$

$$\rightarrow 2 = 3 * a + b, \tag{3.12}$$

$Sequential\_substitute(f(x) = a * x + b, \ [x = 3, \ f(x) = 2])$

$$\rightarrow f(3) = 3 * a + b. \tag{3.13}$$

The operations (3.12) and (3.13) show that sequential substitution is dependent on the order of the substitutions. (Implementation: Maple (mws), Mathematica (nb), MuPAD (mnb).) □

Example 3.33. Consider the three polynomials

$$u(x) = x^2 + x + 2, \qquad v(x) = x^2 + 3 * x - 7, \qquad w(x) = x^2 - 5 * x + 4.$$

In Figure 3.26 we give an MPL dialogue that obtains the functional compositions $u(v(w(x)))$ and $u(w(v(x)))$. Since the composition operation is a not commutative, we use sequential substitution to determine the order of the compositions. □

Definition 3.34. *Let u be an expression and let S be the set of equations*

$$S = \{t_1 = r_1, t_2 = r_2, \ldots, t_n = r_n\},$$

where the targets $t_1, t_2, \ldots, t_n$ are distinct. The concurrent structural substitution operator has the form

$$Concurrent\_substitute(u, S).$$

The operator returns a new expression defined in the following way: recursively search through the expression tree of u and compare each complete sub-expression v to each of the (distinct) targets $t_1, t_2, \ldots, t_n$. If v is identical to some t_i, substitute the corresponding replacement r_i for v.

Since each complete sub-expression of u is identical to at most one target, the order of the substitutions is not significant, and so concurrent substitution is defined in terms of a set S rather than a list.

<1> $u := x^2 + x + 2;$

$$\rightarrow \quad u := x^2 + x + 2$$

<2> $v := x^2 + 3 * x - 7;$

$$\rightarrow \quad v := x^2 + 3 * x - 7;$$

<3> $w := x^2 - 5 * x + 4;$

$$\rightarrow \quad w := x^2 - 5x + 4$$

<4> $Algebraic\_expand(Sequential\_substitute(u, [x = v, x = w]));$

$$\rightarrow \quad x^8 - 20x^7 + 172x^6 - 830x^5 + 2439x^4 - 4390x^3 + 4573x^2 - 2365x + 464$$

<5> $Algebraic\_expand(Sequential\_substitute(u, [x = w, x = v]));$

$$\rightarrow \quad x^8 + 12x^7 + 16x^6 - 234x^5 - 407x^4 + 2202x^3 + 1479x^2 - 10089x + 7834$$

Figure 3.26. An MPL dialogue that obtains a composition of polynomials using sequential substitution. (Implementation: Maple (mws), Mathematica (nb), MuPAD (mnb).)

Example 3.35.

$Concurrent\_substitute((a + b) * x, \ \{a + b = x + c, \ x = d\}) \rightarrow (x + c) * d.$

In this case, the complete sub-expression $a + b$ is replaced by $x + c$ and the complete sub-expression x is replaced by d. Notice since the replacement $x + c$ is not part of the original expression, its x is not replaced by d. If this additional substitution is intended, it is obtained with

$Sequential\_substitute((a + b) * x, [a + b = x + c, \ x = d]) \rightarrow (d + c) * d.$

Another example is

$Concurrent\_substitute(f(x) = a * x + b, \ \{x = 3, \ f(x) = 2\}) \rightarrow 2 = 3 * a + b.$

In this case, the substitution $x = 3$ does not affect the substitution $f(x) = 2$ as it does with sequential substitution. (Compare this with Expression (3.13) where the order of the substitutions affects the result). (Implementation: Maple (mws), Mathematica (nb), MuPAD (mnb).) $\qquad\square$

Most computer algebra systems allow some form of multiple structural substitution (see Figure 3.22). In Exercise 10 on page 195 we describe MPL procedures for these operators.

Since structural substitution is obtained by simply comparing a target expression to the complete sub-expressions of an expression, it cannot obtain all substitutions which occur in symbolic calculations. For more information about algorithms for general substitution operations based on polynomial division, the reader may consult Cohen [24], Sections 4.1 and 6.2.

Exercises

1. For each of the following, give the set of complete sub-expressions of the *automatically simplified* form of the expression:

   (a) $a * b/c$.

   (b) $(a + b) + (c - d)$.

   (c) $1/(2 * x)$.

   (d) $((x - y) * y/2)^2$.

   (e) $x = \dfrac{-b + (b^2 - 4 * a * c)^{1/2}}{2 * a}$.

2. (a) Explain why the operation *Free_of*$(a * (b/c), b/c)$ returns **true**.

   (b) Explore the capacity of the *Free_of* operator in a CAS (see Figure 3.22). Does the operator have the same capacity as the *Free_of* operator in the text?

   (c) One extension of the *Free_of* operator is to allow a target to be a function name or an algebraic operator. Experiment with a CAS to see if the *Free_of* operator in that system has this capability.

3. (a) Explore the capacity of the substitution operator in a CAS (see Figure 3.22). Does the operator have the same capacity as the *Substitute* operator in the text?

   (b) One extension of the *Substitute* operator is to allow a target to be an algebraic operator or a function name. For example, in this case

   $$Substitute(a + b, " + " = " * ") \rightarrow a * b.$$

   Can a CAS do this with its substitution operator?

   (c) Perform each of the following substitutions with a single application of the *Substitute* operator. (In each case, it is necessary to find a "clever" substitution.)

   i. Replace $a * b$ by x in $a * b * c$ to get $x * c$.

   ii. Replace $u + 1$ by x in $(u + 1)^2 + u + 1$ to get $x^2 + x$.

   iii. Replace $a + b$ by 1 in $a * (a + b) + b$ to get 1.

(d) Is it possible to replace all occurrences of the tan function in an expression with its representation in terms of sin and cos with a single application of the *Substitute* operator? For example, is it possible to obtain the transformation $\tan(x) + \tan(y) \to \sin(x)/\cos(x) + \sin(y)/\cos(y)$ with a single substitution? (Don't use a multiple substitution here.)

4. Can the *Solve* operator in a CAS solve an equation for a complete sub-expression? Can the *Solve* operator solve an equation for an expression that is not a complete sub-expression? In Maple use `solve`, in Mathematica use `Solve`, and in MuPAD use `solve`.

5. Evaluate each of the following:

   (a) *Sequential_substitute*$(x * (x + y), [x = 2, \; x + y = 3])$.

   (b) *Concurrent_substitute*$(x * (x + y), \{x = 2, \; x + y = 3\})$.

   (c) *Sequential_substitute*$(x + y^2, [x = y, \; y = x])$.

   (d) *Concurrent_substitute*$(x + y^2, \{x = y, \; y = x\})$.

   (e) *Sequential_substitute*$(a + b + c, [a = b, b = c, c = a])$.

   (f) *Concurrent_substitute*$(a + b + c, \{a = b, b = c, c = a\})$.

6. (a) Let u be an algebraic expression. Give a sequence of statements that gives a new expression with each occurrence of x replaced by y and each occurrence of y replaced by x. For example, $x^2 + 2 * y$ is transformed to $y^2 + 2 * x$. (Don't use a multiple substitution here.)

   (b) Is it possible to do the operation in part (a) with a single statement that involves a multiple substitution?

7. Let u be a mathematical expression and suppose t_1, r_1, t_2, r_2 are distinct symbols. Prove or disprove:

$$Sequential\_substitute(u, [t_1 = r_1, t_2 = r_2]) =$$
$$Sequential\_substitute(u, [t_2 = r_2, t_1 = r_1]).$$

Further Reading

3.1 Recursive Definitions and Algorithms. Recursion for algorithms is discussed in Maurer and Ralston [65]. An interesting popular account of recursion is found in Hofstadter [48].

3.2 Expression Trees and Primitive Operations. Expression trees in a conventional programming context are discussed in most books on data structures and algorithms. For example, see Weiss [98]. Expression trees in the Mathematica system are discussed in Wolfram [102].

4

Elementary Mathematical Algorithms

In this chapter we extend the concept of a mathematical algorithm to include function and procedure definitions, decision structures, and iteration structures. In Section 4.1 we discuss the general concept of a mathematical algorithm and examine some properties of mathematical operators that are used in an algorithm. In Section 4.2 we describe the basic programming structures that are used in MPL and give examples of procedures that use these structures. Finally, in the case study in Section 4.3 we describe a more involved algorithm that finds the solution of some first order ordinary differential equations.

4.1 Mathematical Algorithms

Broadly speaking, a *mathematical algorithm* is a step by step process for solving a mathematical problem that is suitable for computer implementation. Although this definition includes much of what is found in mathematics texts, it is too broad to be useful in practice. We are primarily interested in those algorithms that can be expressed in terms of a computer program using the operators and programming structures available in CAS languages.

Properties of an Algorithm

Computer scientists are quite explicit about the properties a process must have to be called an algorithm. Ideally, a mathematical algorithm should have the following properties.

1. *Each step in the algorithm is precisely defined.*

2. *Each step in the algorithm is **effective** which means it is sufficiently basic so that it can be performed with finite computational resources (time and memory).*

   For example, the operation of multiplying two rational numbers is effective while the operation of multiplying two (mathematical) real numbers that are represented by infinite decimals is not.

3. *The algorithm terminates in a finite number of steps for an appropriate class of input expressions.*

Computer algebra programming differs from conventional programming because the programs contain statements that mimic the symbolic manipulations that are done with pencil and paper calculations. While many of these operations are conceptually well-defined, they are not always algorithmically well-defined in either a theoretical or practical sense. For example, suppose that an algorithm requires the solution of an equation $f(x) = 0$. The algorithm may fail because it is impossible to find a solution for $f(x) = 0$ in terms of a specific class of functions or simply because of the limitations of a CAS's operator for solving equations. One way to resolve the problem is to restrict the algorithm's input to expressions where all operations in the algorithm are well-defined and produce meaningful results. In many cases, however, this is not practical because a description of the valid input would be too involved to be useful.

In some instances, subtle differences in the evaluation process or the actions of operators may cause implementations of an algorithm to perform differently in various computer algebra systems. For example, in Section 7.2 we give a procedure *Simplify_trig(u)* that can verify a large class of trigonometric identities. Implementations of the algorithm in Maple, Mathematica, and MuPAD obtain the simplification

$$(\cos(x) + \sin(x))^4 + (\cos(x) - \sin(x))^4 + \cos(4x) - 3 \to 0.$$

On the other hand, while the Maple and Mathematica implementations obtain the simplification

$$\sin^3(x) + \cos^3\left(x + \frac{\pi}{6}\right) - \sin^3\left(x + \frac{\pi}{3}\right) + \frac{3\sin(3x)}{4} \to 0,$$

the MuPAD implementation obtains

$$\rightarrow -\frac{\sin(x-y)}{2} - \frac{\sin(-x+y)}{2}. \tag{4.1}$$

This discrepancy is explained by the observation that in both Maple and Mathematica the automatic simplification process transforms

$$\sin(-x+y) \rightarrow -\sin(x-y), \tag{4.2}$$

which simplifies the expression (4.1) to zero, while the automatic simplification process in MuPAD does not obtain the transformation (4.2).

If we were to strictly adhere to the formal requirements for an algorithm, we would severely restrict the range of problems that would be attempted in a computer algebra context. Therefore, in describing mathematical algorithms, we take a middle ground between the computer scientist's need for precision and the mathematical scientist's need for practical approaches for solving a problem. In this spirit, we try as much as possible to adhere to the guidelines set down by computer scientists, but also accept that for some input, the theoretical or practical limitations of an operation may cause the algorithm to return an inappropriate result in some instances.

Mathematical Operators in Algorithms

Large computer algebra systems contain more than a thousand mathematical operators in a wide variety of areas. For the algorithms and exercises in this book, we use only a small subset of these operators that perform the basic operations from arithmetic, algebra, trigonometry, calculus, elementary logic, and set theory.

The mathematical operators that are utilized in MPL algorithms are listed below. Some of these operators have been defined informally in previous chapters, and some additional ones are described below.

Algebraic Operators. These are $+$, $-$, $*$, $/$, $\wedge$, and $!$.

Relational and Logical Operators. The relational operators are $=$, $<$, $\leq$, $>$, $\geq$, and $\neq$, and the logical operators are **and**, **or**, and **not**. (See Section 2.1 and Figure 2.6 on page 37.)

Set Operators. These are $\cup$, $\cap$, $\sim$, and $\in$. (See Section 2.4 and Figure 2.20 on page 70.)

List Operators. These are *First*, *Rest*, *Adjoin*, *Join*, *Reverse*, *Delete*, and $\in$. (See page 72 and Figure 2.23 on page 73.)

Primitive Structure Operators. These include the structural operators *Kind, Operand, Number_of_operands,* and *Construct.* (See pages 104-106 and Figure 3.18 on page 106.)

Structure-based Operators. These include the structure-based operators *Free_of, Substitute, Sequential_substitute,* and *Concurrent_substitute* which are based on the tree structure of an expression. (See pages 110-115 and Figure 3.22 on page 111.)

Integer Operators. These operators perform the basic operations on integers. For integers a and $b \neq 0$, using integer division, we obtain a unique quotient q and remainder r with $0 \leq r \leq |b| - 1$, such that $a = q \cdot b + r$. The following operators obtain q and r:

$$Iquot(a, b) \rightarrow q, \qquad Irem(a, b) \rightarrow r.$$

In addition, the operator *Integer_gcd*(a, b) obtains the greatest common divisor of a and b. For further discussion of these operators, see Cohen [24], Section 2.1. The corresponding operators in Maple, Mathematica, and Mu-PAD are given in Figure 4.1. (Implementation: Maple (mws), Mathematica (nb), MuPAD (mnb).)

Calculus Operators. These include the operators *Limit, Derivative,* and *Integral.* (See page 34 and Figure 2.4 on page 35.)

Solution Operators. These are the operator *Solve* that obtains the solutions to some polynomial equations, some systems of polynomial equations, and some algebraic and trigonometric equations, and the operator *Solve_ode* that obtains the solutions to some ordinary differential equations. (See page 34 and Figure 2.4 on page 35.)

Structure Operators for Polynomials. These are the operators *Degree* and *Coefficient.* (See page 63 and Figure 2.4 on page 35.)

Algebraic Manipulation Operators for Polynomials. These include the operators *Factor* and *Algebraic_expand.* (See page 34 and Figure 2.4 on page 35.)

Structure Operators for Rational Expressions. A rational expression is defined as a quotient $u = p/q$ where p and q are polynomials. Two important structural operators for rational expressions are

$$Numerator(u) \rightarrow p, \quad Denominator(u) \rightarrow q.$$

For example,

$$Numerator(x/(x+1)) \to x, \quad Denominator(x^2 + 4\,x) \to 1.$$

The last expression shows that a polynomial is considered a rational expression with denominator 1. The *Numerator* and *Denominator* operators are described in greater detail on page 260, and the corresponding operators in Maple, Mathematica, and MuPAD are given in Figure 4.1. (Implementation: Maple (mws), Mathematica (nb), MuPAD (mnb).)

Simplification Operators. Simplification is such an involved process that it cannot be adequately described in a brief space. For now we utilize two simplification operators. The first one is automatic simplification which is part of the evaluation process described in Sections 2.2 and 3.2. The second one is *Rational_simplify* which transforms an algebraic expression to the form of a rational expression with no common factors in the numerator and denominator.

Example 4.1.

$$Rational\_simplify(1/a + 1/b) \to \frac{a+b}{a\,b},$$

$$Rational\_simplify\left(\frac{x^2 - 1}{x - 1}\right) \to x + 1,$$

$$Rational\_simplify\left(\frac{1}{1/a + c/(ab)} + \frac{abc + ac^2}{(b+c)^2}\right) \to a.$$

(Implementation: Maple (mws), Mathematica (nb), MuPAD (mnb).)   □

The corresponding operators in Maple, Mathematica, and MuPAD are given in Figure 4.1. For more detail on the *Rational_simplify* operator, consult Cohen [24], Section 6.3.

Numerical Operators. The operator *Absolute_value(u)* obtains the absolute value of u. The operator *Decimal(u)* transforms numerical subexpressions of an expression to a decimal format. For example, $Decimal(x + 1/2) \to x + .5$. (See page 34 and Figure 2.4 on page 35.)

Most of these operators and many others are described in greater detail in later chapters.

MPL	Maple	Mathematica	MuPAD
$Iquot(a, b)$	$\mathtt{iquo}(a, b)$	$\mathtt{Quotient}[a, b]$	$\mathtt{iquo}(a, b)$
$Irem(a, b)$	$\mathtt{irem}(a, b)$	$\mathtt{Mod}[a, b]$	$\mathtt{irem}(a, b)$
$Integer\_gcd(a, b)$	$\mathtt{igcd}(a, b)$	$\mathtt{GCD}[a, b]$	$\mathtt{igcd}(a, b)$
$Numerator(u)$	$\mathtt{numer}(u)$	$\mathtt{Numerator}(u)$	$\mathtt{numer}(u)$
$Denominator(u)$	$\mathtt{denom}(u)$	$\mathtt{Denominator}(u)$	$\mathtt{denom}(u)$
$Rational\_simplify(u)$	$\mathtt{normal}(u)$	$\mathtt{Together}(u)$	$\mathtt{normal}(u)$

Figure 4.1. The operators in Maple, Mathematica, and MuPAD that correspond most closely to the MPL operators that are introduced in this section.

Operator Selection

It often happens that a CAS has a number of mathematical operators that can perform a mathematical operation. For example, suppose that a step in a program requires simplifications similar to

$$x^2 - 1 - (x + 1)(x - 1) \to 0. \tag{4.3}$$

In Figure 4.2 we give a Mathematica dialogue that shows three commands that can obtain this simplification. First, the Expand operator, which

---

$In[1] := \mathtt{u = x\char`^2 - 1 - (x + 1) * (x - 1)}$

$Out[1] = -1 + x^2 - (-1 + x)(1 + x)$

$In[2] := \mathtt{Expand[u]}$

$Out[2] = 0$

$In[3] := \mathtt{Together[u]}$

$Out[3] = 0$

$In[4] := \mathtt{Simplify[u]}$

$Out[4] = 0$

Figure 4.2. A Mathematica dialogue that shows a number of mathematical operators that perform an algebraic simplification. (Implementation: Maple (mws), Mathematica (nb), MuPAD (mnb).)

is Mathematica's version of MPL's *Algebraic_expand* operator, applies the distributive law to products and positive integer powers in an algebraic expression. Next, the `Together` operator, which is similar to MPL's *Rational_simplify* operator, performs algebraic expansion as well as more involved operations such as the cancellation of common factors in the numerator and denominator of a rational expression. Finally, the `Simplify` operator is a general purpose simplification operator that applies a large number of algebraic and trigonometric simplification rules to an expression. Similar choices are available in both Maple and MuPAD.

When selecting mathematical operators, to obtain simpler and more efficient programs, we subscribe to the following *minimal power principle*:

> *Always use the least powerful mathematical operator that performs a given mathematical operation.*

For example, if we know that our program will only encounter simplifications similar to Expression (4.3), the CAS's version of the *Algebraic_expand* operator is the most appropriate one to use.

Finally, there are situations where it is clearly inappropriate to use a particular operator in a program. For example, in Section 4.3 we describe a program that finds the solutions to some first order differential equations. It goes with out saying that a CAS's analogue of the *Solve_ode* operator should not be used in this program.

Semantic Capacity of Mathematical Operators

The capability of a mathematical operator can vary from system to system (sometimes dramatically) and may change significantly when a new version of a system is introduced. For example, most computer algebra systems have the capability to compute the limit of a function or an infinite sequence. In mathematics, the limit operation is used in many different contexts, some very concrete and some very abstract. For example:

$$\lim_{x \to \infty} \frac{x^2}{e^x} = 0, \tag{4.4}$$

$$\lim_{x \to \infty} \frac{x^n}{e^x} = 0, \quad (n \text{ an unassigned symbol}), \tag{4.5}$$

$$\lim_{\Delta x \to 0} \frac{f(x + \Delta x) - f(x)}{\Delta x} = \frac{df}{dx}, \tag{4.6}$$

$$\lim_{n \to \infty} \sum_{j=1}^{n} \frac{f(j/n)}{n} = \int_0^1 f(x) \, dx, \tag{4.7}$$

$$\lim_{n \to \infty} r^n = \begin{cases} 0 & -1 < r < 1, \\ 1 & r = 1, \\ \infty & r > 1, \\ \text{undefined} & r < -1. \end{cases} \tag{4.8}$$

For which limit operations should we expect a CAS to obtain a correct result? (Implementation: Maple (mws), Mathematica (nb), MuPAD (mnb).)

To create programs in a CAS language, we must have a clear idea about the capabilities of its mathematical operators. We use the term *semantic capacity* (or just *capacity*) to refer to the mathematical capabilities of an operator. Since the algorithms for mathematical operations can be quite involved, it is often difficult to describe semantic capacity in a simple way. Nevertheless, the concept is an important one even if it cannot be described precisely in some instances. In practice, a useful approach is simply to experiment with a CAS to see what an operator can do. (See Exercise 1 on page 22, Exercise 2 on page 22, and the exercises at the end of this section.)

The following two concepts, *properly posed operations* and *simplification context*, describe some aspects of operator capacity that are useful for understanding the capacity of an operator.

Properly Posed Operations

Informally speaking, a mathematical operation is *properly posed* if all the information needed to perform the operation in an unambiguous manner is available. If an operation is not properly posed, a CAS may not perform the operation, may request additional information, or may return a result that is only correct in some contexts.

In Figure 4.3, we give a Maple dialogue that illustrates three examples of operations that are not properly posed and show how they are handled by this system. At the first prompt, Maple evaluates the indefinite integral

$$\int x^n dx$$

where n is a symbol. This statement is improperly posed because the result depends on whether or not $n \neq -1$ which is unknown at this point. Notice that Maple assumes that $n \neq -1$ and returns the form of the integral for this case.

At the next prompt, we assign to u a second order differential equation with the unassigned symbols a, b, and c, and at the third prompt ask Maple to solve the differential equation. This statement is improperly posed because the form of the solution depends on the value of $b^2 - 4\,a\,c$.

```
> int(x^n, x);
```

$$\frac{x^{(n+1)}}{n+1}$$

```
> u := a*diff(y(x),x,x) + b*diff(y(x),x) + c*y(x) = 0;
```

$$u := a\left(\frac{\partial^2}{\partial x^2}y(x)\right) + b\left(\frac{\partial}{\partial x}y(x)\right) + c\,y(x) = 0$$

```
> dsolve(u, y(x));
```

$$y(x) = \_C1\,e^{\left(-1/2\frac{(b-\sqrt{b^2-4\,a\,c})\,x}{a}\right)} + \_C2\,e^{\left(-1/2\frac{(b+\sqrt{b^2-4\,a\,c})\,x}{a}\right)}$$

```
> w := exp(-s*t)*sin(t);
```

$$w := e^{(-s\,t)}\,\sin(t)$$

```
> int(w, t = 0..infinity);
```

Definite integration: Can't determine if the integral is convergent.
Need to know the sign of $--> $ s
Will now try indefinite integration and then take limits.

$$\lim_{t\to\infty} -\frac{e^{(-s\,t)}\,\cos(t) + s\,e^{(-s\,t)}\,\cos(t) - 1}{s^2+1}$$

```
> assume(s>0);
> normal(int(w, t = 0..infinity));
```

$$\frac{1}{s\text{-}^2+1}$$

Figure 4.3. A Maple interactive dialogue that demonstrates statements that are
improperly posed. (Implementation: Maple (mws), Mathematica (nb), MuPAD
(mnb).)

In this case, Maple returns the form of the solution when $b^2 - 4\,a\,c > 0$,
although when $b^2 - 4\,a\,c = 0$, the correct form is

$$y(x) = c_1\,e^{-b/2a\,x} + c_2\,x\,e^{-b/2a\,x}.$$

At the next two prompts, we ask Maple to evaluate the improper integral

$$\int_0^\infty \exp(-s\,t)\,\sin(t)\,dt.$$

This integral is not properly posed because the convergence of the integral depends on the sign of s. When $s > 0$, the integral converges, and otherwise it diverges. Observe that following the fifth prompt, Maple displays a message indicating that it can't evaluate the integral because it doesn't know the sign of s and then returns an unevaluated limit. At the next prompt, we use Maple's `assume` command to assign the positive property to the symbol s, and at the last prompt we reevaluate the integral and simplify the result with Maple's `normal` command[1].

Similar results are obtained with Mathematica and MuPAD for all three examples.

The question of when an operation is properly posed is an important aspect of operator capacity. A CAS will often make assumptions about the nature of variables in an expression, which means the result returned by an operator may not be correct in all contexts. This can be particularly troubling in an involved program when one of these exceptional situations occurs early in the calculations, remains undetected, and contaminates later calculations. Unfortunately, many mathematical operations that involve general expressions with arbitrary symbols are not properly posed (Exercise 1), and if we try to avoid these situations at all costs our programs will be unnecessarily complicated.

Simplification Context

For efficiency reasons, it is unreasonable to expect a CAS to apply all its simplification rules during the course of a computation. The designer of a CAS must choose which simplification rules are appropriate for a particular operator. We use the term *simplification context* to refer to those simplification rules that are applied during the evaluation of a mathematical operator. The simplification context often determines the form of the output of an operator and in some cases determines whether or not a CAS can even correctly perform an operation.

For example, consider the Maple interactive dialogue in Figure 4.4. At the first prompt, u is assigned a polynomial in x in factored form, and at the second prompt, we ask Maple to obtain the degree of u in x. In this case algebraic expansion (with respect to the symbol x) is part of the

---

[1] Observe that the symbol s in the output of the **normal** command is followed by a tilde (~). The Maple system includes this symbol to indicate that s has been given a property.

```
> u := (x+1)*(x+2);
```

$$u := (x + 1)(x + 2)$$

```
> degree(u,x);
```

$$2$$

```
> v := (y^2-1-(y+1)*(y-1))*x^2+x+1;
```

$$v := (y^2 - 1 - (y + 1)(y - 1))\, x^2 + x + 1$$

```
> degree(v,x);
```

$$2$$

Figure 4.4. A Maple interactive dialogue that demonstrates a simplification context. (Implementation: Maple (mws), Mathematica (nb) MuPAD (mnb).)

simplification context of the **degree** operator which returns the value 2. At the third prompt, v is assigned a polynomial in x with one coefficient that is a polynomial in y (in unexpanded form), and at the next prompt, we ask Maple to obtain the degree of v in x. Notice that the value 2 is returned even though the coefficient of x^2 simplifies to 0. For this system, expansion with respect to the auxiliary symbol y is apparently not part of the simplification context of the **degree** operator. On the other hand, both Mathematica's **Exponent** operator and MuPAD's **degree** operator evaluate the degree of v to 1.

Figure 4.5 shows how the simplification context of the numerator operator can vary from system to system. First, for the expression $(a\,x + b\,x)/c$, Maple returns the numerator in a factored form, while Mathematica returns an expanded form. Next, consider the expression $1/a + 1/b$. In Maple the terms in the sum are combined over a common denominator, and $a + b$ is returned as the numerator. On the other hand, in Mathematica the terms in the sum are not combined, and the entire expression is returned as the numerator.

Simplification context is a rather loosely defined concept. For example, does it refer to the simplification rules that are applied before, during, or after an operation? In addition for some operators a simplification rule

---

```
> u := (a*x+b*x)/c;
```

$$u := \frac{a\,x + b\,x}{c}$$

```
> numer(u);
```

$$(a + b)\,x$$

```
> u := 1/a+1/b;
```

$$u := \frac{1}{a} + \frac{1}{b}$$

```
> numer(u);
```

$$a + b$$

(a) Maple.

$In[1] := \ u = (a * x + b * x)/c$

$Out[1] = \dfrac{a\,x + b\,x}{c}$

$In[2] := \ \texttt{Numerator[u]}$

$Out[2] = \ a\,x + b\,x$

$In[3] := \ u \ = \ 1/a + 1/b$

$Out[3] = \ \dfrac{1}{a} + \dfrac{1}{b}$

$In[4] := \ \texttt{Numerator[u]}$

$Out[4] = \ \dfrac{1}{a} + \dfrac{1}{b}$

(b) Mathematica.

Figure 4.5. Interactive dialogues in Maple and Mathematica that show different simplification contexts of the numerator operation. (Implementation: Maple (mws), Mathematica (nb), MuPAD (mnb).)

may be applied in some situations while not in others or may even depend on other options or settings used in a session. Nevertheless, the concept is an important aspect of operator capacity and serves as a warning that unwarranted assumptions about the actions of an operator may cause a program to fail.

Exercises.

1. Explain why the following operations are not properly posed without additional information about the arbitrary symbols that appear in the expressions. Implement each operation in a CAS. Is the solution obtained correct for all values of the arbitrary symbols?

   (a) $\displaystyle\int \frac{dx}{x^2 + 2\,a\,x + 1}$.

   (b) $\displaystyle\int \sin(n\,x)\,\sin(m\,x)\,dx$.

   (c) Solve the differential equation $\dfrac{dy^2}{dx^2} - y = \exp(a\,x)$.

   (d) $\displaystyle\int_0^1 \frac{1}{x^n}\,dx$.

   Use $\texttt{int}$ and $\texttt{dsolve}$ in Maple, $\texttt{Integrate}$ and $\texttt{DSolve}$ in Mathematica, and $\texttt{int}$ and $\texttt{ode}$ (with $\texttt{solve}$) in MuPAD.

2. Experiment with a CAS to determine the simplification context of the following operators.

   (a) The *Coefficient* operator. For example, can the operator obtain coefficients if the input polynomial is not in expanded form? Is the result returned in expanded form? Is rational simplification part of the simplification context? (Use $\texttt{coeff}$ in Maple and MuPAD and $\texttt{Coefficient}$ in Mathematica.)

   (b) The *Solve* operator. For example, can this operator determine that the quadratic equation $(a^2 - 1 - (a+1)(a-1))\,x^2 + x - 1 = 0$ is really a linear equation and return the solution $x = 1$? How about the equation $(\sin^2(a) + \cos^2(a) - 1)\,x^2 + x - 1 = 0$? Is rational simplification applied before and/or after a solution to the equation is found? (Use $\texttt{solve}$ in Maple and MuPAD and $\texttt{Solve}$ in Mathematica.)

3. Is rational simplification part of the simplification context of the differentiation operator in a CAS? (Use $\texttt{diff}$ in Maple and MuPAD and $\texttt{D}$ in Mathematica.)

4. Describe the semantic capacity of the factor operation in a CAS. The examples in Exercise 1, page 22 are useful for this exercise. (Use $\texttt{factor}$ in Maple and MuPAD and $\texttt{Factor}$ in Mathematica.)

5. Some CAS software has the capability to differentiate and integrate expressions with undefined functions. In this exercise we ask you to explore the capacity of differentiation and integration operators in a CAS to handle undefined functions.

    (a) Apply the differentiation operator to the following expressions:

    $$u(x), \quad \frac{du(x)}{dx} \quad u(x)v(x), \quad u(x)/v(x), \tag{4.9}$$

    $$\sin(u(x)), \quad u(v(x)), \quad x^3 \frac{du(x)}{dx} - 3x^2 u(x) + x^2/2.$$

    Does the CAS obtain the results you expect? (Use `diff` in Maple and MuPAD and `D` Mathematica.)

    (b) Suppose that you are given the derivatives of the expressions in (4.9). Can a CAS integrate these derivatives to obtain the expressions in (4.9) (up to a constant)? (Use `int` in Maple and MuPAD and `Integrate` in Mathematica.)

6. An important aspect of semantic capacity is what an operator does when it is unable to perform the operation. Experiment with a CAS to determine what each of the of the following operators does in this situation.

    (a) *Degree.*

    (b) *Solve.*

    (c) *Integral.*

    (d) *Solve_ode.*

    For example, what does *Degree*(u, x) operator return when the input expression u is not a polynomial in x? (In Maple use `degree, solve, int`, and `dsolve`; in Mathematica use `Exponent, Solve, Integrate`, and `DSolve`; and in MuPAD use `degree, solve, int`, and `ode` (with `solve`).)

4.2   MPL's Algorithmic Language

In this section we describe the basic language structures that are used in MPL to control the flow in an algorithm.

Function Definitions

In ordinary mathematical discourse, the statement, "let $f(x) = x^2 + 4$," defines a computational scheme and does not perform a computation. A computation occurs when the function is invoked with a statement such as $f(2) \to 8$. In MPL, a function definition is used to mimic this operation.

In MPL, a *function definition* has the form

$$f(x_1, \ldots, x_l) \overset{\text{function}}{:=} u,$$

where f is the *function name*, $x_1, \ldots, x_l$ is a sequence of symbols called the *formal parameters*, and u is a mathematical expression. As with ordinary mathematical notation, a function is invoked with an expression of the form

$$f(a_1, \ldots, a_l), \qquad (4.10)$$

where $a_1, \ldots, a_l$ is a sequence of mathematical expressions called the *actual parameters*. When this expression is evaluated, each a_i is evaluated and substituted for the corresponding x_i in u, and then u is evaluated, and the resulting expression is returned as the evaluated form of (4.10).

Example 4.2. Consider the function definition

$$f(x) \overset{\text{function}}{:=} x^2 + 4.$$

The function is invoked with an expression such as $f(2)$. When this statement is evaluated, the actual parameter 2 replaces formal parameter x in $x^2 + 4$, and $f(2) \to 8$. $\qquad \square$

Example 4.3. Consider the function definition

$$T(y, x) \overset{\text{function}}{:=} Derivative(y, x) + y.$$

The function is invoked with an expression such as $T(\sin(t) + t^2, \ t)$. When this statement is evaluated, the actual parameters $\sin(t) + t^2$ and t are substituted for the formal parameters y and x, and we obtain

$$T(\sin(t) + t^2, \ t) \to \cos(t) + 2t + \sin(t) + t^2. \qquad \square$$

In Figure 4.6 we give function definitions in Maple, Mathematica, and MuPAD that implement the MPL definitions in Examples 4.2 and 4.3.

Procedure Definitions

MPL procedures extend the function concept to *mathematical operators* that are defined by a sequence of statements. The general form of a procedure is given in Figure 4.7. The first line of the procedure gives the *procedure name* and a sequence of formal parameters. The **Input** section contains each of the formal parameters x_i along with a brief description

```
f := x -> x^2+4;

T := (y,x) -> diff(y,x)+y;
```

(a) Maple.

```
f[x_] := x^2 + 4

T[y_, x_] := D[y, x] + y
```

(b) Mathematica.

```
f := x -> x^2+4;

T := (y,x) -> diff(y,x)+y;
```

(c) MuPAD.

Figure 4.6. Function definitions in Maple, Mathematica, and MuPAD that correspond to the MPL definitions in Examples 4.2 and 4.3. (Implementation: Maple (mws), Mathematica (nb), MuPAD (mnb).)

of the type of expression that replaces it when the procedure is invoked. MPL procedures always return a mathematical expression as output, and the **Output** section contains a brief description of this expression.

The **Local Variables** section contains a sequence of *local variables* that are known and used only by the procedure. The formal parameters and the local variables make up the *local environment* of a procedure. In a real CAS, each time a procedure is invoked, the variables in this environment are given storage locations in the computer's memory, and when the procedure terminates, these locations are released back to the system.

The statements between the delimiters **Begin** and **End** represent the *body* or the executable statements of the procedure. Each statement S_j is either a mathematical expression, an assignment statement, or a decision or iteration structure both of which are defined later in this section.

A procedure is invoked like a function with an expression of the form (4.10). When the procedure is invoked, each actual parameter a_i is evaluated and then substituted for the corresponding formal parameter x_i, after

---

Procedure $f(x_1, \ldots, x_l)$;
Input
    x_1 : description of input to x_1;
    $\vdots$

    x_l : description of input to x_l;
Output
    description of output;
Local Variables
    $v_1, \ldots, v_m$;
Begin
    S_1;
    S_2;
    $\vdots$

    S_{n-1};
    S_n
End

Figure 4.7. The general form of an MPL procedure.

which each statement S_j in the body is evaluated. In most cases, at least one of the S_j includes a *return statement* that has the form

$$Return(u),$$

where u is a mathematical expression. When this statement is encountered, three actions occur: first, the procedure immediately terminates; second, the evaluated form of u is returned as the evaluated form of Expression (4.10); and finally, control is transferred back to the statement that invoked the procedure. If a *Return* statement is not included, the actions are similar, but now the evaluated form of the last statement S_n is returned by the procedure. We always include a *Return* statement to emphasize what is returned by the procedure.

Example 4.4. We illustrate this concept by defining a procedure that obtains the equation of a tangent line to a function $f(x)$ at the point $x = a$. Recall that the expression for the tangent line is given by

$$\frac{df}{dx}(a)(x - a) + f(a). \tag{4.11}$$

The procedure definition in Figure 4.8 is an algorithmic view of what is done to obtain this expression in expanded form. We invoke the procedure

Procedure *Tangent_line*(f, x, a);
Input
    f : an algebraic expression (formula for a mathematical function);
    x : a symbol (independent variable);
    a : an algebraic expression (point of tangency);
Output
    an algebraic expression that is the formula for the tangent line;
Local Variables
    deriv, m, line;
Begin
1   *deriv* := *Derivative*(f, x);
2   *m* := *Substitute*$(deriv, x = a)$;
3   *line* := *Algebraic_expand*$(m * (x - a) + Substitute(f, x = a))$;
4   *Return*$(line)$
End

Figure 4.8. An MPL procedure that obtains the formula for a tangent line.

with an expression such as

$$Tangent\_line(1/z, z, 3). \tag{4.12}$$

When this expression is evaluated, the three actual parameters $1/z$, z, and 3 are substituted for the corresponding formal parameters f, x, and a, and then the statements in the procedure are evaluated:

$$deriv := Derivative(1/z, \ z) \rightarrow -1/z^2,$$
$$m := Substitute(-1/z^2, \ z = 3) \rightarrow -1/9,$$
$$line := Algebraic\_expand\big((-1/9) * (z - 3) + Substitute(1/z, \ z = 3)\big)$$
$$\rightarrow (-1/9)\, z + 2/3.$$

Therefore
$$Tangent\_line(1/z, z, 3) \rightarrow (-1/9)\, z + 2/3.$$

When we invoked the procedure in Expression (4.12), for clarity we intentionally chose names for mathematical symbols that were different from the formal parameter names of the procedure. There is no reason, however, to restrict the actual parameters in this way. For example, the procedure can also be invoked with

$$Tangent\_line(1/x, x, 3) \rightarrow (-1/9)\, x + 2/3. \tag{4.13}$$

Keep in mind, however, that the actual parameter x in Statement (4.13) and the formal parameter x in the procedure declaration

$$\textbf{Procedure } \textit{Tangent\_line}(f, x, a) \qquad\qquad (4.14)$$

are different symbols even though they have the same name. When Statement (4.13) is evaluated, each actual parameter is substituted for the corresponding formal parameter which means that f is replaced by $1/x$, the formal parameter x in (4.14) by the actual parameter x in (4.13), and a by 3. Therefore, the differentiation at line 1 is

$$\textit{Derivative}(1/x, x) \to -1/x^2,$$

where the x that appears here is the one in (4.13). Similar comments apply to the other statements in the procedure. $\square$

Maple, Mathematica, and MuPAD provide procedures that operate as described above. In Figures 4.9 and 4.10 we give implementations of *Tangent_line* in these languages.

Global Symbols.   A symbol that appears in a function or a procedure that is not a formal parameter or a local variable is called a *global symbol*. Global symbols, which are accessible to both the interactive mode and other functions and procedures, provide another way to pass data to and from a procedure without using the formal parameters or a *Return* statement.

For a simple example, consider a modification of the *Tangent_line* procedure in which the variable *deriv* has been removed from the local section and therefore is considered global. In this case, after evaluating Statement (4.12) the global variable *deriv* has the value $-1/z^2$ which can now be used by other functions, procedures, or the interactive mode.

In our MPL procedures, global symbols are used primarily to return information about the status of an operation. For example, in Figure 4.14 on page 144 we give a procedure that tries to determine if a mathematical function is even or odd. The procedure returns one of the global symbols *Even* or *Odd* when the input expression is even or odd, or the global symbol *Unknown* when the procedure cannot determine the property.

Use of Local Variables in MPL.   Procedures provide a way to isolate part of a computation so that programming variables in the local environment do not conflict with variables with the same name in other functions, procedures, or the interactive mode. However, in some systems local variables

```
Tangent_line := proc(f,x,a)
#Input
# f: an algebraic expression (formula for a mathematical function)
# x: a symbol (independent variable)
# a: an algebraic expression (point of tangency)
#Output
# an algebraic expression that is the formula for the tangent line
local
  deriv,m,line;
deriv := diff(f,x);
m := subs(x=a,deriv);
line := expand(m*(x-a)+subs(x=a,f));
RETURN(line)
end:
```

(a) Maple.

```
TangentLine[f_,x_,a_] := Module[
(*Input
  f: an algebraic expression (formula for a mathematical function)
  x: a symbol (independent variable)
  a: an algebraic expression (point of tangency)
Output
  an algebraic expression that is the formula for the tangent line
Local*)
  {deriv,m,line},
deriv  = D[f,x];
m = ReplaceAll[deriv,x->a];
line = Expand[m*(x-a)+ ReplaceAll[f,x->a]];
Return[line]
]
```

(b) Mathematica.

Figure 4.9. Implementations of the MPL procedure in Figure 4.8 in Maple and Mathematica. (Implementation: Maple (txt), Mathematica (txt).)

```
Tangent_line := proc(f,x,a)
/*Input
   f: an algebraic expression (formula for a mathematical function)
   x: a symbol (independent variable)
   a: an algebraic expression (point of tangency)
Output
   an algebraic expression that is the formula for the tangent line
*/
local
   deriv,m,line;
begin
deriv := diff(f,x);
m := subs(deriv,x=a);
line := expand(m*(x-a)+subs(f,x=a));
return(line)
end_proc:
```

Figure 4.10. A MuPAD implementation of the MPL procedure in Figure 4.8. (Implementation: MuPAD (txt).)

can also act as mathematical symbols in an expression, and when this happens, name conflicts can occur that are not encountered with conventional programming languages.

For example, suppose that a symbol x is declared local in a procedure, and suppose that it is used as a mathematical symbol in an expression that is returned by a procedure. When this happens, does x lose some of its local characteristics? For example, when this x is returned to the interactive mode, is it the same as a mathematical symbol x used elsewhere in the interactive mode?

In Figure 4.11 we show how this situation is handled by the Maple system. At the first prompt, we define a procedure F(a) that returns the expression $a*x^2$ with the local x. At the second prompt, we call on the procedure and assign the output to u. At the third prompt, we differentiate u with respect to x and obtain what appears to be an incorrect result. The problem here is the local x in the procedure and the x in the diff command are different symbols even though they have the same displayed name in the interactive mode.

In Mathematica, local variables in a procedure can act as mathematical symbols, although a procedure similar to the Maple procedure in Figure 4.11 returns the expression $a\,x^2$ with the symbol name x replaced by another system-generated name.

```
> F := proc(a)
  local x;
  RETURN(a*x^2)
  end:
> u := F(3);
```

$$u := 3\,x^2$$

```
> diff(u,x);
```

$$0$$

Figure 4.11. A Maple dialogue in which a local mathematical symbol is returned from a procedure. (Implementation: Maple (mws), Mathematica (nb), MuPAD (mnb).)

The MuPAD system avoids this situation altogether by not permitting unassigned local variables in a procedure to act as mathematical symbols. To avoid conflicts of this sort and to provide a system-independent programming style, we follow MuPAD's lead and adopt the following convention:

> *In MPL procedures, an unassigned local variable cannot appear as a symbol in a mathematical expression.*

In other words, in MPL procedures local variables can only act as programming variables and must be assigned before they appear in a mathematical expression. In situations where a procedure requires a local mathematical symbol, we either pass the symbol through the parameter list or use a global symbol.

Use of Formal Parameters in MPL. In conventional programming languages, a procedure's formal parameters can be used both to transmit data to and from a procedure and as local variables. The situation with CAS languages is more involved, however, because the actual parameters in a procedure call can be mathematical expressions as well as variables. Because of this, the language mechanism that is used to bind the formal parameters with the actual parameters can be rather involved and can vary from system to system. For this reason, the use of formal parameters for anything but the transmission of data into a procedure is system dependent. Since our goal is to present a system-independent programming style, we adopt the following convention:

Formal parameters in MPL procedures are used only to transmit data into a procedure and not as local variables or to return data from a procedure.

When we need to return more than one expression from a procedure, we return a list of expressions.

Decision Structures

Decision structures provide a way to control the flow in an algorithm. MPL provides three decision structures. The simplest one is the **if** structure which has the general form shown in Figure 4.12-(a). The expression

---

if *condition* **then**
  T_1;
  T_2;
  $\vdots$
  T_m;

(a) The **if** structure.

if *condition* **then**
  T_1;
  T_2;
  $\vdots$
  T_m
else
  F_1;
  F_2;
  $\vdots$
  F_n;

(b) The **if-else** structure.

Figure 4.12. The general form of the MPL **if** and **if-else** decision structures.

condition is a logical (or relational) expression that evaluates to one of the logical constants **true** or **false**. Each T_i is either a mathematical expression, an assignment statement, another decision structure, or an iteration structure (described below).

The **if** structure usually operates in the following way: when *condition* evaluates to **true**, the indented[2] statements $T_1, T_2, \ldots, T_m$ are evaluated, and when *condition* evaluates to **false** these statements are skipped. The exception to this scheme arises when the **if** statement is included in a procedure, and one of the indented statements includes a *Return*. In this case, when *condition* is **true**, the statements controlled by the **if** are evaluated until the *Return* is encountered, at which point the procedure terminates, and the evaluated form of the argument to *Return* is returned by the procedure. This exception also applies to the other decision and iteration structures described below.

A more general decision structure is the **if-else** structure which allows for two alternatives. It has the general form[3] shown in Figure 4.12-(b). When the expression *condition* evaluates to **true**, the statements $T_1, T_2, \ldots, T_m$ are evaluated, and when *condition* evaluates to **false**, the statements $F_1, F_2, \ldots, F_n$ are evaluated.

Example 4.5. Here is a simple example of an **if-else** structure:

$$\begin{aligned}
&\textbf{if}\ \ 0 \leq x\ \textbf{and}\ x \leq 1\ \textbf{then} \\
&\quad f := x^2 + 4 \\
&\textbf{else} \\
&\quad f := x^2 - 1;
\end{aligned} \tag{4.15}$$

(Implementation: Maple (mws), Mathematica (nb), MuPAD (mnb).)   □

The most general MPL decision structure is the *multi-branch* decision structure which allows for a sequence of conditions. It has the general form shown in Figure 4.13. In this generality, the structure contains zero or more **elseif** sections and an optional **else** section. Upon evaluation, the logical expressions $condition_1, condition_2, \ldots$ are evaluated in sequence. If $condition_i$ is the first one that evaluates to **true**, then the statements in that section $S_{i1}, \ldots, S_{im_i}$ are evaluated while all the other statements are

---

[2] Some computer algebra languages require a termination symbol (such as **end_if**, **fi**, or **]**) to indicate the extent of statements controlled by the **if** structure. In MPL, these statements are indicated by indentation without a termination symbol.

[3] As is common practice in some programming languages, in MPL we omit the semicolon at the end of a statement that precedes an **else**, an **elseif** (defined below), and an **End**.

> **if** $condition_1$ **then**
> $\quad S_{11};$
> $\quad S_{12};$
> $\qquad \vdots$
> $\quad S_{1m_1}$
> **elseif** $condition_2$ **then**
> $\quad S_{21};$
> $\quad S_{22};$
> $\qquad \vdots$
> $\quad S_{2m_2}$
>
> $\qquad \vdots$
>
> **elseif** $condition_n$ **then**
> $\quad S_{n1};$
> $\quad S_{n2};$
> $\qquad \vdots$
> $\quad S_{nm_n}$
> **else**
> $\quad F_1;$
> $\quad F_2;$
> $\qquad \vdots$
> $\quad F_r;$

Figure 4.13. The MPL multi-branch structure that provides for a sequence of alternatives.

skipped. If none of the tests evaluate to **true**, the statements in the **else** section (if included) are evaluated.

All computer algebra languages provide **if** structures and **if-else** structures, and some languages provide a version of the multi-branch decision structure[4].

The procedure in the next example utilizes a multi-branch structure.

---

[4] In Maple and MuPAD, use the `if` statement to implement MPL's **if**, **if-else**, and multi-branch structures. In Mathematica, use the `If` statement to implement MPL's **if** and **if-else** structures and the `Which` statement to implement MPL's multi-branch structure.

Example 4.6. Recall that a mathematical function $u(x)$ is even if

$$u(x) - u(-x) = 0$$

and odd if

$$u(x) + u(-x) = 0.$$

For example, $u(x) = x^2 - 1$ is even, $u(x) = x^3$ is odd, while $u(x) = x^2 + x^3$ is neither even nor odd.

A procedure that tries to determine if an algebraic expression u is even or odd is given in Figure 4.14. The procedure is interesting for both what it can do and what it cannot do. Observe that the procedure operates in the simplification context of automatic simplification, and in this context it can determine the nature (even or odd) of the first two examples given above.

Notice that when the procedure is unable to determine that u is even or odd, it returns the symbol *Unknown*, rather than a symbol indicating that the expression is neither even nor odd. We do this because automatic simplification applied at lines 2 and 4 may not simplify an expression to zero even though the expression simplifies to zero in a mathematical sense. For example, suppose that u is the even expression $(x + 1)(x - 1)$, and

---

Procedure *Even_odd(u, x)*;
Input
    u : an algebraic expression;
    x : a symbol;
Output
    one of the global symbols *Even*, *Odd*, or *Unknown*;
Local Variables
    v;
Begin
1    $v := Substitute(u, x = -x)$;
2   **if** $u - v = 0$ **then**
3   *Return(Even)*
4   **elseif** $u + v = 0$ **then**
5   *Return(Odd)*
6   **else**
7   *Return(Unknown)*
End

Figure 4.14. An MPL procedure that attempts to determine if u is even or odd. (Implementation: Maple (txt), Mathematica (txt), MuPAD (txt).)

let's assume that algebraic expansion is not included in automatic simplification[5]. In this case, v is the expression $(-x + 1)(-x - 1)$, and $u - v$ is the expression $(x + 1)(x - 1) - (-x + 1)(-x - 1)$, which does not simplify to 0 with automatic simplification. Although we can remedy this by applying the *Algebraic_expand* operator at lines 2 and 4, there are other expressions that are not handled in this simplification context. For example, $1/(x - 1) - 1/(x + 1)$ is even, but this cannot be determined by algebraic expansion and automatic simplification. In this case, rational simplification (with *Rational_simplify*) is required at lines 2 and 4. But then, $\sin(x/(x + 1)) + \sin(x/(x - 1))$ is even, but this is not handled by rational simplification.

While it is possible to increase the simplification power at lines 2 and 4 to handle all of the above expressions, it is theoretically impossible to increase the simplification power to a level that the procedure can always determine if an algebraic expression is even or odd[6].   □

Iteration Structures

MPL contains two iteration structures that allow for repeated evaluation of a sequence of statements. The first iteration structure is the **while** structure which has the general form

$$\begin{aligned} &\textbf{while} \quad condition \ \textbf{do} \\ &\quad S_1; \\ &\quad S_2; \\ &\quad \vdots \\ &\quad S_n; \end{aligned} \tag{4.16}$$

where *condition* is a logical (or relational) expression. This structure is evaluated by first evaluating *condition*, and if it is to **true**, the indented statements $S_1, S_2, \ldots, S_m$ are evaluated. Once this is done, the process repeats, and again if the logical *condition* is **true**, the indented statements are evaluated. The process continues in this way checking if *condition* is **true** and if so, evaluating the indented statements. On the other hand once

---

[5] In Maple, Mathematica, and MuPAD, algebraic expansion is not part of automatic simplification.

[6] The problem to determine if an expression simplifies to 0 is known as the *zero equivalence problem*. D. Richardson has shown that for the class of algebraic expressions constructed with rational numbers, the symbol x, the real numbers π and $\ln(2)$, the sin, exp, and absolute value functions, and sums, products, and powers with integer exponents, it is impossible to give an algorithm that can always determine if an expression simplifies to 0 (see Richardson [84]).

condition evaluates to **false**, the indented statements are not evaluated, and the structure terminates.

Example 4.7. The sum of the first $n + 1$ terms of a Taylor series for a function $u(x)$ about $x = a$ is given by

$$\sum_{i=0}^{n} \frac{u^{(i)}(a)}{i!}(x - a)^i \qquad (4.17)$$

where $u^{(i)}$ is the ith derivative of $u(x)$, and $u^{(0)} = u(x)$. When n is a non-negative integer, the sum (4.17) is obtained with the following MPL statements:

```
1    i := 1;
2    s := Substitute(u, x = a);
3    while i ≤ n do
4        u := Derivative(u, x);
5        s := s + Substitute(u, x = a)/i! * (x − a)^i;
6        i := i + 1;
```

The substitution in line 2 initializes s to $u^{(0)}(a) = u(a)$, and each traversal through the **while** loop adds one additional term of the Taylor series to s and increases the counter i by 1. Eventually $i = n+1$, and so the condition $i \leq n$ is **false**, and the **while** structure terminates.

For example, if $u = \sin(x)$, $n = 3$, and $a = 0$, after executing the loop we obtain $s = x - x^3/6$. (Implementation: Maple (mws), Mathematica (nb), MuPAD (mnb).) $\qquad \square$

The second iteration structure is the **for** structure which has the general form

$$\textbf{for } i := \textit{start} \textbf{ to } \textit{finish} \textbf{ do}$$
$$S_1; \qquad\qquad\qquad\qquad (4.18)$$
$$S_2;$$
$$\vdots$$
$$S_n;$$

where i is a variable and *start* and *finish* are expressions that evaluate to integer values. When $start \leq finish$, the indented statements are evaluated for each integer value of $i = start, start + 1, \ldots, finish$. If $start > finish$, the indented statements are not evaluated[7]

Example 4.8. The sum of the first $n + 1$ terms of the Taylor series can also be obtained using a **for** structure:

$$
\begin{array}{ll}
1 & s := Substitute(u, x = a); \\
2 & \textbf{for } i := 1 \textbf{ to } n \textbf{ do} \\
3 & \quad u := Derivative(u, x); \\
4 & \quad s := s + Substitute(u, x = a)/i! * (x - a)^i;
\end{array}
$$

(Implementation: Maple (mws), Mathematica (nb), MuPAD (mnb).)   □

All computer algebra languages provide iteration structures similar to **while** and **for**[8].

Evaluation of Logical Expressions. In MPL, the value (**true** or **false**) of a logical expression with main operator **and** or main operator **or** is obtained by evaluating each of the operators in a left to right manner until the value of the entire expression is determined. In some cases this value is obtained without evaluating all the operands of the logical expression. For example, consider the following decision structure which tests if n is a positive integer:

$$\textbf{if } Kind(n) = \textbf{integer and } n > 0 \textbf{ then} \tag{4.19}$$

$$\vdots$$

Observe that the second relational expression only evaluates to **true** or **false** when n has a numerical value. When n is not an integer, however, the value of the entire logical expression (**false**) is determined by the test $Kind(n) = \textbf{integer}$, and there is no need to evaluate the expression $n > 0$.

Most computer algebra systems evaluate logical expressions in decision and iteration structures in a similar way[9].

---

[7] Some of our procedures contain **For** loops that include a *Return* statement. (For example, see lines 5-6 in the procedure *Polynomial_sv* in Figure 6.2 on page 218.) In this case, we intend that both the loop and the current procedure terminate when the *Return* is encountered, and that the value returned by the procedure is the value of the operand of the *Return* statement. The **for** statements in both Maple and MuPAD work in this way. However, in Mathematica, a **Return** in a **For** statement will only work in this way if the upper limit contains a relational operator (e.g., `i<=N`). (Implementation: Mathematica (nb).)

[8] In Maple and MuPAD, use the **while** and **for** statements. In Mathematica, use the **While** and **For** statements.

[9] Maple, Mathematica, and MuPAD use this approach to evaluate logical expressions in decision and iteration structures.

The procedure in the next example uses the concepts described in this section.

Example 4.9. It is often necessary to separate the operands of a product into two classes, those that depend on an expression (say x) and those that do not. For example, this operation is needed when we use the linear property of the integral to move the factors of a product that do not depend on the integration variable x outside of the integral sign:

$$\int \frac{c\,x\sin(x)}{2}dx = \frac{c}{2}\int x\,\sin(x)dx. \qquad (4.20)$$

A procedure *Separate_factors* that performs the separation operation is given in Figure 4.15. The procedure takes two algebraic expressions u

---

Procedure *Separate_factors*(u, x);
Input
      u, x : algebraic expressions;
Output
      a list with two algebraic expressions;
Local Variables
      $f, free\_of\_part, dependent\_part, i$;
Begin
1   **if** $Kind(u) = " * "$ **then**
2      $free\_of\_part := 1$;
3      $dependent\_part := 1$;
4   **for** $i := 1$ **to** $Number\_of\_operands(u)$ **do**
5        $f := Operand(u, i)$;
6   **if** $Free\_of(f, x)$ **then**
7          $free\_of\_part := f * free\_of\_part$
8   **else**
9          $dependent\_part := f * dependent\_part$;
10      $Return([free\_of\_part, dependent\_part])$
11   **else**
12   **if** $Free\_of(u, x)$ **then**
13        $Return([u, 1])$
14   **else**
15        $Return([1, u])$
End

Figure 4.15. An MPL procedure that separates factors in a product that depend on x from those that do not. (Implementation: Maple (txt), Mathematica (txt), MuPAD (txt).)

and x as input and returns a two-element list. The first member of the list contains the product of the factors of u that are free of x, while the second member contains the product of the remaining factors. If there are no factors in a category, the integer 1 is returned for that category.

The procedure can be applied to both products and non-products. When u is a product, the *Free_of* operator is applied to each factor which is then placed in the appropriate category (lines 6-9). When u is not a product, it is reasonable to apply *Free_of* to the entire expression which is then placed in the appropriate category (lines 12-15). The procedure is invoked with an expression such as

$$Separate\_factors\left(\frac{c\,x\,\sin(x)}{2},\,x\right) \to [c/2,\,x\,\sin(x)]. \qquad \square$$

Comparison of the MPL and CAS Languages

In Chapters 2, 3, and this chapter we have introduced the main elements of the MPL algorithmic language. The description includes the following elements.

1. *The MPL mathematical operators.* A summary of these operators is given on pages 121-123 of this chapter. In later chapters many of these operators are described in greater detail and many others are introduced.

2. *A description of the evaluation process including automatic simplification.* All calculations in our programs are done in the context of automatic simplification. Automatic simplification is described in Chapters 2 and 3. For a more detailed discussion of automatic simplification consult Cohen [24], Chapter 3.

3. *The structure of mathematical expressions.* Mathematical expressions are the data objects of computer algebra. The form of these expressions in the context of automatic simplification is described in Chapter 3.

4. *The MPL algorithmic structures.* Functions, procedures, decision structures, and iteration structures are described in this section, and a few additional ones are described in later chapters.

Although MPL is similar to real CAS languages, it models only a small subset of these languages. Large CAS languages contain over 1000 mathematical operators and other language features that provide greater mathematical power, facilitate the programming process, and enhance the computational efficiency of programs.

There is, however, much to be gained from MPL's simplicity. MPL's algorithms can be implemented (usually with only minor modifications) in many real CAS languages using only the basic operations of these languages. In fact, many mathematical operations can be formulated in terms of the analogues of MPL's primitive operators (*Kind*, *Operand*, etc.) or in terms of other operators that are defined in terms of these primitive operators.

Exercises

Unless otherwise noted, each of the functions and procedures in the exercises should be expressed in terms of a CAS's version of the mathematical operators given on pages 121-123.

1. Consider the function $f(x) = \dfrac{1}{1-x}$.

   (a) Show that $f(f(f(x))) = x$ with pencil and paper.

   (b) Define this function using a function definition in a CAS language.

   (c) Use a CAS to show that $f(f(f(x))) = x$.

2. (a) The curvature of a function $f(x)$ is given by

$$k(x) = \frac{|f''(x)|}{(1 + (f'(x))^2)^{3/2}}.$$

   Give a procedure *Curvature*(f, x) that computes the curvature of an algebraic expression f at x.

   (b) Apply the *Curvature* operator to the function

$$f(x) = \sqrt{4 - x^2}.$$

   Since this function represents the positive semicircle of radius 2, the curvature result simplifies to the value $1/2$. Can you obtain this simplification with a CAS?

3. Let u be an equation that represents a straight line in x and y, and let p be a two-element list of rational numbers that represents the coordinates of a point.

   (a) Give a procedure *Perpendicular_line*(u, x, y, p) that returns the equation of a line perpendicular to u that passes through the point p. Be sure to include the cases for horizontal and vertical lines. For example,

$$Perpendicular\_line(2\,x + 3\,y = 4,\ x,\ y,\ [1, 2]) \rightarrow y - 2 = (3/2)\,(x - 1).$$

(b) Give a procedure

$$Distance\_point\_line(u, x, y, p)$$

that returns the shortest distance from the point p to the line u. For example,

$$Distance\_point\_line(2\,x + 3\,y = 4,\ x,\ y,\ [1, 2]) \rightarrow (4/13)\,\sqrt{13}.$$

4. Let u be a mathematical expression. Give a procedure $Operand\_list(u)$ that returns the operands of a compound expression in a list. (The operands in the list should be in the same order as the operands in u.) If u is not a compound expression, return the global symbol **Undefined**. If u is a list, return u. For example,

$$Operand\_list(a + b + c) \rightarrow [a, b, c].$$

5. Let u be an equation of the form $f = g$ where f and g are polynomials in x with coefficients that are rational numbers such that $f - g$ has degree ≤ 2. Give a procedure $Solve\_quadratic(u, x)$ that finds the roots of the equation $f = g$. Be sure to check if $f - g$ is a constant, linear, or quadratic polynomial. Do not use a CAS's solve operator in this problem.

6. Let S be a set of polynomials in x. Give a procedure $Find\_min\_deg(S, x)$ that returns a polynomial of smallest degree in S. If $S = \emptyset$, return the global symbol **Undefined**.

7. The *set product* of sets A and B is the set of all lists $[x, y]$ where $x \in A$ and $y \in B$. This set is represented by $A \times B$. If either $A = \emptyset$ or $B = \emptyset$, then, by definition, $A \times B = \emptyset$. Give a procedure $Set\_product(A, B)$ that returns $A \times B$. For example,

$$Set\_product(\{a, b\},\ \{c, d\}) \rightarrow \{[a, c],\ [a, d],\ [b, c],\ [b, d]\}.$$

8. Let x be a symbol, and let u be a polynomial in x with rational number coefficients. Give a procedure $Linear\_factors(u, x)$ that returns the product of the linear factors of u. If u has no linear factors, return 1. Use the factor operator in a CAS to obtain the factorization of u. For example,

$$Linear\_factors(x^2 + x, x) \rightarrow x\,(x + 1), \qquad Linear\_factors(x^3 + 1, x) \rightarrow x + 1,$$

$$Linear\_factors(x^2 + 1, x) \rightarrow 1, \quad Linear\_factors(x^2 + 2\,x + 1, x) \rightarrow (x + 1)^2.$$

9. Let u be a polynomial in x and y with rational number coefficients. A polynomial u is *symmetric* if it is not changed when the variables x and y are interchanged. For example, the polynomial $u = x^2 + 2\,x\,y + y^2$ is symmetric. Give a procedure $Symmetric(u, x, y)$ that returns **true** if u is symmetric and **false** otherwise.

10. Let L be a list. Give a procedure $Remove\_duplicates(L)$ that returns a new list with all members that are identical to a previous member of the list removed from u. For example, $Remove\_duplicates([a, b, c, a, c]) \rightarrow [a, b, c]$.

11. Let u be an algebraic expression. The *numerical coefficient part* of u is defined in the following way:

   (a) If u is a rational number, the numerical coefficient part of u is u.

   (b) If u is a product, the numerical coefficient part is the operand of u that is a rational number. If this operand does not exist, then the numerical coefficient part is 1.

   (c) If u is any other type of expression, then the numerical coefficient part is 1.

   Let n be the numerical coefficient part of an expression. Give a procedure *Numerical_coefficient*(u) that returns a two-element list $[n, u/n]$. For example, *Numerical_coefficient*$(2/3\, x\, \sin(x)) \to [2/3, x\, \sin(x)]$.

12. Let u be an algebraic expression. Give a procedure *Separate_sin_cos*(u) that returns a two-element list $[r, s]$ that is defined using the following rules.

   (a) If u is a product, then s is the product of the operands of u that are sines, cosines, or positive integer powers of sines and cosines, and r is the product of the remaining operands of u. (If there are no operands in a category, return 1 for that category.)

   (b) If u is a sine, cosine, or a positive integer power of a sine or cosine, then $s = u$ and $r = 1$.

   (c) In all other cases, $r = u$ and $s = 1$.

   For example,

   $$Separate\_sin\_cos(3\, \sin(x)\, \cos(y)) \to [3, \sin(x)\, \cos(y)],$$

   $$Separate\_sin\_cos(1 + \sin(x)) \to [1 + \sin(x), 1].$$

   This procedure is used in the procedure *Contract_trig_rules* in Figure 7.7, page 297.

13. Let u be an algebraic expression, and let x and y be symbols. Give a procedure

   $$Separate\_variables(u, x, y)$$

   that determines if an expression u can be factored in the form $u = p \cdot q$, where p is free of y, and q is free of x. Use the factor operator in a CAS to obtain the factorization of u. If u can be factored in this form, return a list $[p, q]$, otherwise return **false**. For example,

   $$Separate\_variables(3\, x\, y + 3\, x, \ x, \ y) \to [3\, x, \ y + 1],$$

   $$Separate\_variables(x + y, \ x, \ y) \to \textbf{false}.$$

   This procedure is used in the procedure *Separable_ode* described in Exercise 5 on page 168.

14. Let $P = [[x_1, y_1], \ldots, [x_{r+1}, y_{r+1}]]$ be a list of 2 element lists, where x_i and y_i are rational numbers. The *Lagrange interpolation polynomial* that passes through these points is given by

$$L(x) = \sum_{i=1}^{r+1} y_i L_i(x).$$

where

$$L_i(x) = \frac{(x - x_1) \cdots (x - x_{i-1})(x - x_{i+1}) \cdots (x - x_{r+1})}{(x_i - x_1) \cdots (x_i - x_{i-1})(x_i - x_{i+1}) \cdots (x_i - x_{r+1})}.$$

Give a procedure *Lagrange_polynomial*(P, x) that returns the polynomial $L(x)$. For example,

$$Lagrange\_polynomial([[1, 1], [2, -1]], \ x) \to -2\,x + 3.$$

15. Let u be an equation that involves x and y, and suppose that it is possible to solve the equation for y as a linear expression in x using algebraic operations such as rational simplification and expansion. Give a procedure $Line(u, x, y)$ that solves the equation for y and returns the result in the form $y = m\,x + b$. Do not use the solve operator in a CAS in this exercise. For example, your procedure should obtain the following transformations:

$$Line\left(\frac{x}{2} + \frac{y}{3} = 1, \ x, \ y\right) \to y = (-3/2)\,x + 3,$$

$$Line\left(\frac{x}{a} = \frac{x + y}{b}, \ x, \ y\right) \to y = \frac{b - a}{a}\,x,$$

$$Line\left(\frac{y/x - 2}{1 - 3/x} = 6, \ x, \ y\right) \to y = 8\,x - 18.$$

16. A Taylor series for a function $u(x, y)$ about the point (a, b) is given by

$$T(x, y) = \sum_{i=0}^{\infty} u_i(x, y)/i!, \tag{4.21}$$

where

$$u_i(x, y) = \sum_{j=0}^{i} \frac{i!}{(i - j)!\,j!} \frac{\partial^i u(a, b)}{\partial x^{i-j} y^j} (x - a)^{i-j} (y - b)^j.$$

For example, for $u = \exp(x)\cos(y)$ and $(a, b) = (0, 0)$, the Taylor series is

$$T(x, y) = 1 + x + (1/2)(x^2 - y^2) + (1/6)(x^3 - x\,y^2) - (1/3)\,x\,y^2 + \cdots$$

Let u, a, and b be algebraic expressions, x and y be symbols, and n a non-negative integer. Give a procedure *Taylor_2*(u, x, y, a, b, n) that obtains the sum of the first $n + 1$ terms of the series (4.21). *Note:* A more efficient procedure is obtained by using the expression $u_{i-1}(x, y)$ to obtain the next expression $u_i(x, y)$.

17. Consider the differential equation and initial condition

$$\frac{dy(x)}{dx} = f(x, y(x)), \qquad y(a) = b. \qquad (4.22)$$

A Taylor series solution to this equation, which has the form

$$y(a) + \frac{dy(x)}{dx}(a)\,(x - a) + \frac{d^2 y(x)}{dx^2}(a)\,\frac{(x - a)^2}{2!} + \cdots,$$

is found in the following way. The constant term in the series is given by
the initial condition in (4.22), and the second term is obtained using the
differential equation in (4.22)

$$\frac{dy(x)}{dx}(a)\,(x - a) = f(a, y(a))\,(x - a).$$

The third term is obtained by differentiating both sides of the differential
equation

$$\frac{d^2 y(x)}{dx^2} = \frac{df(x, y(x))}{dx},$$

and which gives

$$\frac{d^2 y(x)}{dx^2}(a)\,\frac{(x - a)^2}{2!} = \left.\frac{df(x, y(x))}{dx}\right|_{x=a}\frac{(x - a)^2}{2!}.$$

The next term in the series is obtained in a similar way with the second
derivative of $f(x, y(x))$. For example, for the differential equation and
initial condition

$$\frac{dy(x)}{dx} = f(x, y(x)) = x^3 + \frac{1}{y(x)} + 3, \qquad y(0) = 2,$$

the first term is $y(0) = 2$, and the second term is $f(0, y(0))\,(x - 0) = 7/2\,x$.
To obtain the third term, we first obtain an expression for the second
derivative using the differential equation

$$\frac{d^2 y(x)}{dx^2} = \frac{df(x, y(x))}{dx} = 3x^2 - \frac{\dfrac{dy(x)}{dx}}{y(x)^2} = 3x^2 - \frac{x^3 + 1/y(x) + 3}{y(x)^2},$$

and then using the substitutions $y(x) = 2$ and $x = 0$ to obtain

$$\frac{d^2 y(x)}{dx^2}(0)\,(x - 0)^2/2! = -7/16\,x^2.$$

In a similar way the fourth term of the series is $35/64\,x^3$.

Let w be a differential equation in the form (4.22), x and y be symbols,
a and b be algebraic expressions, and n be a non-negative integer. Give a
procedure

$$Taylor\_ode(w, x, y, a, b, n)$$

that obtains the sum of the first $n + 1$ terms of the Taylor series solution to the differential equation. *Note*: Under suitable conditions on $f(x, y)$ the Taylor series converges to $y(x)$ for x in an interval about $x = a$. In this case the polynomial obtained by *Taylor_ode* is an approximation to the true solution to the differential equation.

18. Consider the two infinite series

$$F = \sum_{n=0}^{\infty} \frac{f_n(t - t_0)}{n!}, \qquad G = \sum_{n=0}^{\infty} \frac{g_n(t - t_0)}{n!}, \qquad (4.23)$$

where the functions $f_n = f_n(t)$ and $g_n = g_n(t)$ are defined by the relations

$$f_n = \frac{df_{n-1}}{dt} - \mu(t)\, g_{n-1}, \qquad (4.24)$$

$$g_n = f_{n-1} + \frac{dg_{n-1}}{dt}, \qquad (4.25)$$

with the initial functions given by

$$f_0 = 1, \qquad (4.26)$$

$$g_0 = 0. \qquad (4.27)$$

The two series in 4.23) are known in astronomy as the F and G series where they are used for orbit calculations.

The computation in this problem is one of the early (1965) applications of computer algebra that used the FORMAC computer algebra system developed at IBM (Bond et al. [11]). In this problem we restrict our attention to the symbolic computation problem associated with the computation of the functions f_n and g_n. Observe that Equation (4.24) contains an undefined function $\mu(t)$, which implies that f_n and g_n also depend on t, and therefore the differentiations in Equations (4.24) and (4.25) make sense. Using the relations in (4.24) and (4.25) and the initial terms (4.26) and (4.27), the next two terms of each sequence are given by

$$f_1 = 0, \quad g_1 = 1, \quad f_2 = -\mu(t), \quad g_2 = 0. \qquad (4.28)$$

For larger values of n, it is customary in astronomical calculations to define two additional functions $\sigma(t)$ and $\epsilon(t)$ and to make the substitutions

$$\frac{d\mu(t)}{dt} = -3\mu(t)\,\sigma(t), \quad \frac{d\sigma(t)}{dt} = \epsilon(t) - 2\sigma(t)^2, \quad \frac{d\epsilon(t)}{dt} = -\sigma(t)\,(\mu(t) + 2\epsilon(t)),$$
$$(4.29)$$

whenever these derivatives appear in the calculations. For example, to compute f_3, we use Equations (4.24), (4.28), and (4.29) to obtain

$$f_3 = \frac{df_2}{dt} - \mu(t)\, g_2 = \frac{d(-\mu(t))}{dt} - \mu(t) \cdot 0 = 3\,\mu(t)\,\sigma(t).$$

In a similar way, we have

$$g_3 = -\mu(t), \quad f_4 = -15\,\mu(t)\,\sigma(t)^2 + 3\,\mu(t)\,\epsilon(t) + \mu(t)^2, \quad g_4 = 6\,\mu(t)\,\sigma(t).$$

For larger values of n, the algebra becomes much more involved, and so this is a good candidate for computer algebra.

(a) Using Equations (4.24), (4.25), (4.26), (4.27), and (4.28), show that

$$\begin{aligned}
f_5 &= 105\,\sigma(t)^3\,\mu(t) - 45\,\mu(t)\,\epsilon(t)\,\sigma(t) - 15\,\sigma(t)\,\mu(t)^2, \\
g_5 &= -45\,\sigma(t)^2\,\mu(t) + 9\,\epsilon(t)\,\mu(t) + \mu(t)^2.
\end{aligned}$$

(b) Let n be a positive integer, and let t be a symbol. Give a procedure $FG(n, t)$ that returns the list $[f_n, g_n]$, where f_n and g_n are expressed in terms of $\mu(t)$, $\sigma(t)$, and $\epsilon(t)$.

4.3   Case Study: First Order Ordinary Differential Equations

In this section we describe an algorithm that finds a solution to some first order differential equations using techniques similar to those found in an elementary differential equations textbook. A first order ordinary differential equation is one in which the highest order derivative is a first derivative. For example,

$$x\,\frac{dy}{dx} + y^2 = x - 1$$

is a first order differential equations, while

$$\frac{d^2y}{dx^2} + y = \sin(x)$$

is not. Although first order differential equations are very difficult to solve in general, there are some specific forms that are solvable.

The solution technique we use involves the method of separation of variables and the method of exact equations using integrating factors. In the next few pages we describe these approaches in enough detail to allow us to formulate our procedures. Additional theory and examples can be found in most differential equations textbooks[10].

---

[10]For example, see Simmons [87], Chapters 1 and 2, Boyce and DiPrima [12], Chapter 2, or Derrick and Grossman [32], Chapter 2.

Separation of Variables

A differential equation that can be expressed in the form

$$\frac{dy}{dx} = f(x)\,g(y) \tag{4.30}$$

is called a *separable* differential equation. In this case, the notation implies that the expression to the right of the equal sign can be factored as a product of an expression that is free of y and one that is free of x. To solve the equation, divide both sides by $g(y)$ and integrate with respect to x

$$\int \frac{1}{g(y)}\frac{dy}{dx}\,dx = \int f(x)\,dx.$$

By the chain rule, this is equivalent to

$$\int \frac{dy}{g(y)} = \int f(x)\,dx.$$

By integrating both sides of this equation, we obtain an implicit solution to the differential equation.

Example 4.10. Consider the differential equation, $\dfrac{dy}{dx} = 2\,x\,y^2$. An implicit solution is given by

$$\int \frac{dy}{y^2} = \int 2x\,dx,$$
$$\frac{-1}{y} = x^2 + C.$$

In this case, by solving for y we obtain an explicit solution

$$y = \frac{-1}{x^2 + C}. \tag{4.31}$$

In most cases, however, it is difficult (or impossible) to express the solution in explicit form. For this reason, our algorithm returns the result in implicit form. □

Exact Differential Equations and Integrating Factors

This technique applies to differential equations that can be transformed to the form

$$M(x,y) + N(x,y)\frac{dy}{dx} = 0. \tag{4.32}$$

Our goal is to find an implicit solution to this equation that has the form

$$g(x, y) = C, \tag{4.33}$$

where C is an arbitrary constant. To obtain a solution algorithm, let's suppose this expression is a solution to Equation (4.32). Considering y as a function of x, differentiating Equation (4.33) with the chain rule gives

$$\frac{dg}{dx} = \frac{\partial g}{\partial x} + \frac{\partial g}{\partial y}\frac{dy}{dx} = 0. \tag{4.34}$$

Comparing this equation to Equation (4.32), we obtain

$$\frac{\partial g}{\partial x} = M(x, y), \qquad \frac{\partial g}{\partial y} = N(x, y), \tag{4.35}$$

and find the solution to Equation (4.32) by solving these two equations for $g(x, y)$.

Example 4.11. Consider the differential equation

$$2x + 3y^2 + (6xy + y^2)\frac{dy}{dx} = 0. \tag{4.36}$$

We find a solution by solving the equations

$$\frac{\partial g}{\partial x} = 2x + 3y^2, \qquad \frac{\partial g}{\partial y} = 6xy + y^2. \tag{4.37}$$

Integrating the first of these equations with respect to x, we obtain

$$g(x, y) = \int \frac{\partial g}{\partial x}\, dx = \int 2x + 3y^2\, dx = x^2 + 3xy^2 + h(y). \tag{4.38}$$

Since this operation inverts the partial differentiation operation, we assume that y is fixed during the integration and obtain a constant of integration $h(y)$ that may depend on y, but is free of x. To find $h(y)$, using Equation (4.38) we differentiate $g(x, y)$ with respect to y,

$$\frac{\partial g(x, y)}{\partial y} = \frac{\partial (x^2 + 3xy^2 + h(y))}{\partial y} = 6xy + h'(y),$$

and compare this result with the second equation in (4.37). Therefore,

$$6xy + h'(y) = 6xy + y^2,$$

which implies $h'(y) = y^2$. Integrating with respect to y, we obtain $h(y) = y^3/3$, and therefore an implicit solution to the differential equation is

$$g(x, y) = x^2 + 3\,x\,y^2 + y^3/3 = C. \qquad (4.39)$$

We can also start the process by integrating the second equation in (4.37) with respect to y:

$$g(x, y) = \int \frac{\partial g}{\partial y}\,dy = \int 6\,x\,y + y^2\,dy = 3\,x\,y^2 + y^3/3 + k(x),$$

where now the constant of integration depends on x. Differentiating this expression with respect to x and comparing the result with the first expression in (4.37), we obtain $k(x) = x^2$, which gives again the solution in Equation (4.39). $\qquad\square$

The next example shows that the method does not always work.

Example 4.12. Consider the differential equation

$$2 + 3\,y/x + (3 + 3\,y^2/x)\frac{dy}{dx} = 0, \qquad x > 0. \qquad (4.40)$$

We try to find a solution by solving the equations

$$\frac{\partial g}{\partial x} = 2 + 3\,y/x, \qquad \frac{\partial g}{\partial y} = 3 + 3\,y^2/x. \qquad (4.41)$$

Integrating the first equation with respect to x, we obtain

$$g(x, y) = \int \frac{\partial g}{\partial x}\,dx = \int (2 + 3\,y/x)\,dx = 2\,x + 3\,y\,\ln(x) + h(y). \qquad (4.42)$$

To find $h(y)$, we differentiate this expression with respect to y

$$\frac{\partial(2\,x + 3\,y\,\ln(x) + h(y))}{\partial y} = 3\,\ln(x) + h'(y)$$

and compare this result with the second equation in (4.41). We obtain $3\,\ln(x) + h'(y) = 3 + 3\,y^2/x$, which implies $h(y)$ is not free of x and so the technique does not work. In addition, if we start the process by first integrating

$$\frac{\partial g}{\partial y} = 3 + 3\,y^2/x$$

with respect to y and then differentiating $g(x, y)$ with respect to x, we find that the constant of integration $k(x)$ is not free of y and so again the technique does not work. $\qquad\square$

As we saw in the last example, for the technique to work the constants of integration ($h(y)$ or $k(x)$) must be free of the other variable (x or y). Equations for which this happens are called *exact* differential equations. There is a simple test that determines if an equation is exact. It can be shown[11] that an equation is exact if and only if

$$\frac{\partial M}{\partial y} = \frac{\partial N}{\partial x}. \qquad (4.43)$$

Using this relation, we can easily check that Equation (4.36) is exact, while Equation (4.40) is not.

When the equation is not exact, it may be possible to transform the equation to one that is exact. We illustrate this in the next example.

Example 4.13. Consider again the differential equation from the last example

$$2 + 3\,y/x + (3 + 3\,y^2/x)\,\frac{dy}{dx} = 0,$$

where $x > 0$. If we multiply both sides of the equation by $u(x, y) = x$, we obtain a new differential equation

$$2\,x + 3\,y + (3\,x + 3\,y^2)\,\frac{dy}{dx} = 0. \qquad (4.44)$$

Since

$$\frac{\partial M}{\partial y} = 3 = \frac{\partial N}{\partial x},$$

Equation (4.44) is exact, and the solution technique for exact equations gives the implicit solution $x^2 + 3\,x\,y + y^3 = C$. $\square$

The expression $u(x, y)$ in the previous example is called an *integrating factor* for the differential equation. Although an integrating factor always exists in theory, it may be very difficult to find in practice[12]. Two cases where simple integrating factors can be found are described in the following theorem.

Theorem 4.14. *Consider the differential equation*

$$M(x, y) + N(x, y)\frac{dy}{dx} = 0.$$

---

[11]See Simmons [87], pp. 51-52, Boyce and DiPrima [12], page 84, Derrick and Grossman [32], page 41.

[12]For example, see Boyce and DiPrima [12], page 87, where it is shown that the integrating factor is a solution to a partial differential equation. Unfortunately, it may be very difficult to solve the partial differential equation.

1. *Let*

$$F = \frac{\frac{\partial M}{\partial y} - \frac{\partial N}{\partial x}}{N}. \tag{4.45}$$

If F is free of y, then an integrating factor is $u = \exp\left(\int F\,dx\right)$.

2. *Let*

$$G = \frac{\frac{\partial N}{\partial x} - \frac{\partial M}{\partial y}}{M}. \tag{4.46}$$

If G is free of x, then an integrating factor is $u = \exp\left(\int G\,dy\right)$.

In either case, $u\,M + u\,N\dfrac{dy}{dx} = 0$ is an exact differential equation.

Example 4.15. Consider again the inexact equation

$$2 + 3\,y/x + (3 + 3\,y^2/x)\frac{dy}{dx} = 0, \qquad x > 0.$$

We have

$$F = \frac{\left(\frac{\partial M}{\partial y} - \frac{\partial N}{\partial x}\right)}{N} = \frac{3/x + 3y^2/x^2}{3 + 3\,y^2/x} = 1/x, \tag{4.47}$$

where the expression on the right is obtained with rational simplification. Since this expression is free of y, $u = \exp(\int 1/x\ dx) = x$ is an integrating factor and we obtain the exact form of Equation (4.44).

On the other hand, since

$$G = \frac{\left(\frac{\partial N}{\partial x} - \frac{\partial M}{\partial y}\right)}{M} = \frac{-3\,y^2/x^2 - 3/x}{2 + 3\,y/x} = \frac{-3\,x - 3\,y^2}{2\,x^2 + 3\,x\,y}$$

is not free of x, this approach does not obtain an integrating factor that is free of x. ☐

The *Solve_ode* Algorithm

An MPL algorithm that attempts to solve a first order differential equation is given in Figures 4.16 and 4.17. The algorithm returns either an implicit solution to the differential equation, which may include some unevaluated integrals, or the global symbol **Fail** if it cannot find a solution using the methods described in this section. The main procedure of the algorithm is

Procedure *Solve_ode*(w, x, y);
Input
    w : a differential equation that can be transformed by
        rational simplification to the form $M + N\dfrac{dy}{dx} = 0$,
        where the derivative $\dfrac{dy}{dx}$ is represented by the function form $d(y, x)$;
    x, y : symbols;
Output
    An implicit solution to the differential equation or the global symbol **Fail**;
Local Variables
    p, M, N, F;
Begin

```
1    p := Transform_ode(w, x, y);
2    M := Operand(p, 1);
3    N := Operand(p, 2);
4    F := Separable_ode(M, N, x, y);
5    if  F = Fail then
6        F := Solve_exact(M, N, x, y)
7    Return(F)
```

End

Procedure *Transform_ode*(w, x, y);
Input
    same as *Solve_ode*;
Output;
    the list $[M, N]$;
Local Variables
    v, n, M, N;
Begin

```
1    v := Rational_simplify(Operand(w, 1) − Operand(w, 2));
2    n := Numerator(v);
3    M := Coefficient(n, d(y, x), 0);
4    N := Coefficient(n, d(y, x), 1);
5    Return([M, N])
```

End

Figure 4.16. The MPL procedures *Solve_ode* and *Transform_ode*. (Implementation: Maple (txt), Mathematica (txt), MuPAD (txt).)

Procedure $Solve\_exact(M, N, x, y)$;
Input
    M, N: algebraic expressions;
    x, y: symbols;
Output
    An implicit solution to the differential equation or the global symbol **Fail**;
Local Variables
    $My, Nx, d, u, F, G, g, h, hp$;
Begin

```
1      if  N = 0 then
2          Return(Fail)
3      elseif  M = 0 then
4          Return(y = C);
5      My := Derivative(M, y);
6      Nx := Derivative(N, x);
7      d := My − Nx;
8      if  d = 0 then
9          u := 1
10     else
11         F := Rational_simplify(d/N);
12         if  Free_of(F, y) then
13             u := exp(Integral(F, x));
14             d := 0
15         else
16             G := Rational_simplify(−d/M);
17             if  Free_of(G, x) then
18                 u := exp(Integral(G, y));
19                 d := 0;
20     if  d = 0 then
21         g := Integral(u ∗ M, x);
22         hp := u ∗ N − Derivative(g, y);
23         h := Integral(hp, y);
24         Return(g + h = C)
25     else
26         Return(Fail)
   End
```

Figure 4.17. The MPL $Solve\_exact$ procedure. (Implementation: Maple (txt), Mathematica (txt), MuPAD (txt).)

Solve_ode(w, x, y), where w is the differential equation with the derivative symbol

$$\frac{dy}{dx}$$

represented by the function form[13] $d(y, x)$.

At line 1, we invoke the *Transform_ode* procedure which does some preliminary manipulation of the equation and returns the list $[M, N]$ with the expressions $M(x, y)$ and $N(x, y)$ in (4.32). This procedure, which is shown in the bottom of Figure 4.16, permits some flexibility in the form of the input equation. For example, by preprocessing the equation with this procedure we can handle equations with forms like

$$1 - (2\,x + 1)\,d(y, x) = 0, \quad 1 - 2\,x\,d(y, x) = -d(y, x),$$

or even

$$1/d(y, x) = 2\,x + 1. \tag{4.48}$$

At line 1 of *Transform_ode* we subtract the right side of the equation from the left side and then simplify this expression using the *Rational_simplify* operator. Next, line 2 selects the numerator of v. For example, if w is given by Equation (4.48), then after executing lines 1 and 2 we have

$$n := 1 - 2\,x\,d(y, x) - d(y, x). \tag{4.49}$$

In lines 3 and 4 we view n as a polynomial in $d(y, x)$ and retrieve M and N by selecting coefficients of this polynomial. For example, for Equation (4.48) the procedure returns $[1, -2\,x - 1]$.

At this point, control is returned to *Solve_ode* which obtains M and N and then calls on *Separable_ode* to find a solution (lines 2, 3, and 4). This procedure attempts to solve

$$\frac{dy}{dx} = -M/N$$

using the separation of variables technique. (The *Separable_ode* procedure is described in Exercise 5.) If this method fails, the *Solve_exact* procedure, which attempts to solve the differential equation using the method of exact equations, is invoked at line 6.

The *Solve_exact* procedure is shown in Figure 4.17. To begin, two simple cases are considered in lines 1-4. First, if $N = 0$, there is no first derivative

---

[13] In Maple and MuPAD we represent the derivative with $d(y, x)$, while in Mathematica we use $d[y, x]$. We use this notation instead of the derivative operator in a CAS because the details of the *Solve_ode* algorithm are somewhat simpler with this representation. This representation for the derivative is also used in Exercise 15 on page 197 and Exercise 15 on page 240.

term in the equation, and so the procedure returns the global symbol **Fail**. Next, if $N \neq 0$ and $M = 0$, the differential equation is equivalent to

$$\frac{dy}{dx} = 0,$$

and so the constant solution is returned. Lines 5 and 6 compute the partial derivatives in Expression (4.45) and line 7 evaluates the difference of these derivatives so we can test if the equation is exact[14]. At line 8, if $d = 0$ the equation is exact and an integrating factor is not required. Therefore, u is assigned the expression 1 at line 9, and control is transferred to line 20. On the other hand, if at line 8 $d \neq 0$, we assume the equation is not exact and compute and test F to determine if there is an integrating factor that is free of y. Notice we apply the *Rational_simplify* operator in line 11 since automatic simplification may not remove the symbol y from F (see Example 4.15 above). The free-of test is done in line 12, and if it is successful we compute the integrating factor in line 13. The assignment in line 14 allows the procedure to proceed with the solution technique in line 20. If the test in line 12 fails we compute and test G to determine if there is an integrating factor that is free of x (lines 18 - 19).

In line 20, if $d = 0$, we apply the method of exact equations (lines 21-23) and return an implicit solution at line 24. If at line 20, $d \neq 0$ an integrating factor has not been found, and so we return the symbol **Fail** in line 26.

Theory versus Practice

In a theoretical sense, a separable equation can be solved using the method of exact equations by expressing Equation (4.30) in the exact form

$$-f(x) + \frac{1}{g(y)}\frac{dy}{dx} = 0.$$

In practice, however, the manipulations in the procedure *Transform_ode* may transform a separable equation in exact form to a non-exact equation that cannot be solved by *Solve_exact*. This point is illustrated in the next example.

Example 4.16. Consider the separable equation

$$\frac{-x}{x+2} + \frac{y}{y+1}\frac{dy}{dx} = 0.$$

---

[14]Observe that d is computed in the context of automatic simplification. Although this context is sufficient when M and N are polynomials in x and y, it is possible to construct equations where additional simplification power is needed.

For this equation, the manipulations in *Transform_ode* obtain $M = -y\,x - x$ and $N = x\,y + 2\,y$ which gives a differential equation in non-exact form. In addition, at lines 11 and 16 in *Solve_exact*, we obtain $F = -(x + y)/(x\,y + 2\,y)$ which is not free of y and $G = -(x + y)/(x\,y + x)$ which is not free of x. Therefore, *Solve_exact* cannot find an integrating factor and so it cannot find a solution to the equation.   □

Unfortunately, there are other (non-separable) exact equations that loose their exactness in *Transform_ode* and cannot be solved with *Solve_exact* (see Exercise 3).

Appraisal of the Algorithm

Given appropriate input, the *Solve_ode* algorithm finds the general solution to many first order differential equations found in textbooks on ordinary differential equations. In addition, another approach for the integrating factor and special techniques for homogeneous equations and Bernoulli equations that extend the capacity of the algorithm are described in Exercises 4, 6, and Exercise 16 on page 241. However, compared to the differential equation solver found in a CAS, the algorithm is quite limited. The operators in these systems include additional techniques for many special forms and other general techniques[15].

In some cases the implicit solution that is found by our algorithm does not describe all solutions to the differential equation.

Example 4.17. Consider the differential equation

$$\frac{dy}{dx} = 2\,x\,y^2 \tag{4.50}$$

given in Example 4.10. Our algorithm finds the solution in the form $-x^2 - 1/y = C$ which has the explicit form $y = -1/(x^2 + C)$. Observe that $y = 0$ is also a solution of the differential equation, but does not fit the general pattern. This solution, which is not found by our algorithm, is called a *singular* solution of the differential equation.   □

In order for the MPL algorithm to produce an appropriate result, the input differential equation must have a form that can be analyzed correctly

---

[15]In the Maple system, to see the methods used by the `dsolve` command, assign `infolevel[dsolve] := 3`. Try this for the differential equation

$$\frac{dy}{dx} = \frac{x + y + 4}{x - y - 6}$$

that cannot be solved by the algorithm in this section (including the additional techniques in the exercises), but which can be solved by Maple.

by the *Transform_ode* procedure. A suitable form is one that can be transformed by the operations in lines 1 and 2 of this procedure to a form where
the actions of the *Coefficient* operator in lines 3 and 4 are well-defined and
able to obtain the entire structure of the equation. If this is not so, the
output of the algorithm may be meaningless. For example, for the equation

$$\left(\frac{dy}{dx}\right)^{1/2} + x = y,$$

the expression $n = (d(y,x)) \wedge (1/2) + x - y$ at line 2 is not a polynomial in
$d(y,x)$ and so the coefficient operations in lines 3 and 4 are undefined. In
addition, for differential equations that contain higher order derivatives or
an integer power of a derivative, the coefficient operations may be defined
but the output is meaningless since the algorithm does not apply to equations that include these forms. It is possible to modify *Transform_ode* so
that is does a more thorough analysis of the input equation to determine
if the equation has an appropriate form (see Exercise 14 on page 240).

Exercises

1. Consider the differential equation

$$(2y - x^2) + (2x - y^2)\frac{dy}{dx} = 0.$$

   Solve the equation using the algorithm in the text.

2. Consider the differential equation $(y - 1/x)\frac{dy}{dx} + y/x^2 = 0$.

   (a) Show that the equation is exact.

   (b) Show that the manipulations in the *Transform_ode* procedure transform the equation to a non-exact equation.

   (c) Show that the *Solve_exact* procedure can find the solution to the new
   equation obtained in part (b) by finding an integrating factor.

3. Consider the differential equation $\dfrac{1}{x^3y^2} + \left(\dfrac{1}{x^2y^3} + 3y\right)\dfrac{dy}{dx} = 0.$

   (a) Show that the equation is exact.

   (b) Show that the manipulations in the *Transform_ode* procedure transform the equation to a non-exact equation.

   (c) Show that the *Solve_ode* procedure is unable to solve this equation
   because it is unable to find an integrating factor for the equation in
   part (b). (However, see Exercise 4.)

4. Let $R = (\partial M/\partial y - \partial N/\partial x)/(N \cdot y - M \cdot x)$ and suppose that R is a function of the product $x\,y$. In this case, it can be shown that for $z = x\,y$,

$$u(x, y) = \exp\left(\int R(z)\,dz\right)$$

is an integrating factor[16]. For example, for the differential equation

$$y + (x + 3x^3 y^4)\frac{dy}{dx} = 0, \quad x > 0, \; y > 0,$$

we have $R = -3/(x\,y) = -3/z$ and $u = 1/(x\,y)^3$. Extend the *Solve_exact* procedure so that it determines when this integrating factor is appropriate and when this is so, uses it to find a solution. Test the procedure on the above equation. *Hint:* Let $S = Substitute(R, x = z/y)$. If R has the proper form, then S is free of y.

5. Give a procedure *Separable_ode*(M, N, x, y) that tries to determine if a differential equation (4.32) can be transformed to the form of Equation (4.30), and, when this is so, obtains an implicit solution using the separable approach. If this technique does not apply, return the global symbol **Fail**. *Hint:* The *Separate_variables* procedure described in Exercise 13 on page 152 is useful in this exercise.

6. A differential equation that can be transformed to the form

$$\frac{dy}{dx} = f(y/x) \tag{4.51}$$

is called a *homogeneous*[17] differential equation. For example, the equation

$$\frac{dy}{dx} = \exp(y/x) + y/x$$

is homogeneous. A homogeneous differential equation can be solved by defining a new variable $z = y/x$ and transforming the differential equation to one in terms of z. Using the relation $y = x\,z$, we have

$$\frac{dy}{dx} = x\frac{dz}{dx} + z$$

and Equation (4.51) becomes

$$\frac{dz}{dx} = (f(z) - z)/x. \tag{4.52}$$

This equation can be solved by separating the variables x and z. Then, we obtain the solution to Equation (4.51) by substituting $z = y/x$ into the solution to Equation (4.52). Give a procedure *Homogeneous*(M, N, x, y)

---

[16] See Simmons [87], Exercise 1 on page 59.

[17] The term *homogeneous* has a number of meanings with regard to differential equations. For example, two different meanings are given in Exercises 6 and 7.

that determines if a first order differential equation (4.32) is homogeneous and, if so, solves the equation using the approach outlined above. If the equation is not homogeneous, return the global symbol **Fail**. *Hint:* First, represent the differential equation in the form

$$\frac{dy}{dx} = -M/N,$$

and let $r = Substitute(f, y = z\,x)$. If r is free-of x, the original equation is homogeneous. Note that the equation

$$\frac{dy}{dx} = \frac{x+y}{x-y}$$

is homogeneous (divide the numerator and denominator by x). In this case $r = (x + x\,z)/(x - x\,z)$, and so we must apply a *Rational_simplify* operator to r to remove the x.

7. Consider the second order linear differential equation

$$a\,\frac{d^2 y}{dx^2} + b\,\frac{dy}{dx} + c\,y = f, \qquad (4.53)$$

where a, b, and c are rational numbers and f is an algebraic expression that is free of y.

(a) If $f = 0$, the equation is called a *homogeneous*[17] equation and two linearly independent solutions to the differential equation y_1 and y_2 are obtained as follows: let $D = b^2 - 4ac$. If $D > 0$, then

$$y_1 = \exp((-b + \sqrt{D})/(2a)x), \quad y_2 = \exp((-b - \sqrt{D})/(2a)x).$$

If $D = 0$, then

$$y_1 = \exp(-b/(2a)x), \quad y_2 = x\,\exp(-b/(2a)x).$$

If $D < 0$, then

$$
\begin{aligned}
y_1 &= \exp(-b/(2\,a)\,x)\,\sin(\sqrt{-D}/(2\,a)\,x), \\
y_2 &= \exp(-b/(2\,a)\,x)\,\cos(\sqrt{-D}/(2\,a)\,x).
\end{aligned}
$$

Give a procedure

$$Homogeneous\_2(a, b, c, x)$$

that returns the list $[y_1, y_2]$.

(b) A particular solution y_p to Equation (4.53) is obtained using the method *variation of parameters*. Using this technique, $y_p = v_1\,y_1 + v_2\,y_2$ where y_1 and y_2 are the two linearly independent solutions to the homogeneous equation (described above) and the derivatives v_1' and v_2' satisfy the linear system

$$v_1'\,y_1 + v_2'\,y_2 = 0, \qquad v_1'\,y_1' + v_2'\,y_2' = f/a.$$

The expressions v_1 and v_2 are obtained from their derivatives by integration. Give a procedure $Variation\_of\_param(y1, y2, f, a, x)$ that obtains y_p.

(c) The general solution to the differential equation is given by

$$y = d\, y_1 + e\, y_2 + y_p, \qquad (4.54)$$

where d and e are symbols that represent arbitrary constants. Give a procedure $Solve\_ode\_2(a, b, c, f, x, y)$ that obtains the general solution to Equation (4.53). You should return the result as an equation $y = u$ where u is the expression on the right side of Equation (4.54). A related operator is considered in Exercise 15 on page 240.

8. See Exercise 14 on page 240 and Exercise 16 on page 241.

Further Reading

4.2 MPL's Algorithmic Language. The Taylor series solution to a differential equation described in Exercise 17 on page 154 is discussed in Zwillinger [109], Section 140. See Sconzo et al. [86] for a discussion of the classical hand calculation of F and G series (see Exercise 18 on page 155) and a summary of the results obtained with a CAS.

4.3 Case Study: Solution of First Order Ordinary Differential Equations. The techniques used in this section are described in Simmons [87], Boyce and DiPrima [12], and Derrick and Grossman [32]. Zwillinger [109] and Murphy [72] describe many techniques for finding analytical solutions to differential equations. Postel and Zimmermann [81] summarizes techniques for solving differential equations in a computer algebra context.

5

Recursive Algorithms

In this chapter we examine how recursion is used to implement algorithms in computer algebra. We begin, in Section 5.1, by describing how a simple recursive procedure is implemented by a CAS. In Section 5.2, we give recursive procedures for a number of operators and describe an approach using transformation rules that provides a simple way to implement some recursive operations. Finally, in Section 5.3 we describe a recursive algorithm for a simple version of the *Integral* operator that utilizes some basic integration rules together with the substitution method.

5.1 A Computational View of Recursion

In Chapter 3 we gave the following recursive definition for the factorial operation:

$$n! = \begin{cases} 1, & \text{if } n = 0, \\ n \cdot (n-1)!, & \text{if } n > 0. \end{cases} \tag{5.1}$$

For $n = 4$, the computation based on this definition (5.1) proceeds as follows:

$$\begin{aligned} 4! = 4(3!) = 4(3(2!)) = 4(3(2(1!))) &= 4(3(2(1(0!)))) \\ &= 4(3(2(1(1)))) \tag{5.2} \\ &= 24. \end{aligned}$$

To perform the calculation, we repeatedly apply (5.1) until $n = 0$ is encountered. Once this point is reached, 0! is replaced by the value 1, and the numerical computation proceeds as indicated by the parentheses in the second line of Equations (5.2).

Procedure   *Rec_fact(n)*;
Input
   n : non-negative integer;
Output
   n!;
Local Variables
   f;
Begin
1   **if** *n* = 0 **then**
2   *f* := 1
3   **else**
4   *f* := *n* ∗ *Rec_fact(n − 1)*
5   *Return(f)*
End

Figure 5.1. An MPL recursive procedure for $n!$. (Implementation: Maple (txt), Mathematica (txt), MuPAD (txt).)

Figure 5.1 shows an MPL recursive procedure that performs this calculation. For the case $n > 0$, the procedure calls on itself (line 4) to perform a "simpler" version of the calculation. A procedure that calls on itself directly (as in this example) or indirectly through a sequence of procedures is called a *recursive procedure*. The case $n = 0$ (lines 1, 2) is called a *termination condition* for the procedure, since it is defined directly and does not require further calls on *Rec_fact*. For each positive integer n, the calculation is eventually reduced to the termination condition which stops the recursion. Each recursive procedure must have one or more termination conditions.

Let's trace the execution of the procedure in response to the evaluation of *Rec_fact*(4) from the interactive mode. When the procedure is invoked, a CAS allocates a block of computer memory that includes storage locations for the local variable f, the input variable n, and the next statement executed by the system once *Rec_fact* is done. The storage allocation for *Rec_fact*(4) (before the calculation in line 4) is shown in Figure 5.2(a). At this point, the local variable f has not been assigned, and the "next statement executed" refers to the interactive mode that invoked the procedure and will display the result once the operation is done.

The actual calculation is done in line 4. But before this can be done, we need the value for *Rec_fact*(3), and this requires another call on the procedure. To invoke *Rec_fact*(3), a CAS again allocates a block of memory

n	f	next statement executed
4		interactive mode

(a) The storage allocation stack for $Rec\_fact(4)$
before calculation on line 4.

n	f	next statement executed
3		Rec_fact, line 4 ($n = 4$ case)
4		interactive mode

(b) The storage allocation stack for $Rec\_fact(3)$
and $Rec\_fact(4)$. The local variable f has not
been assigned a value in either block.

n	f	next statement executed
0	1	Rec_fact, line 4 ($n = 1$ case)
1		Rec_fact, line 4 ($n = 2$ case)
2		Rec_fact, line 4 ($n = 3$ case)
3		Rec_fact, line 4 ($n = 4$ case)
4		interactive mode

(c) The storage allocation stack for
the sequence of Rec_fact procedure
calls before the recursion unwinds.

Figure 5.2. The storage allocation stack for the procedure Rec_fact at various
points in the computation of 4!.

to store the information associated with this procedure call. Figure 5.2(b)
illustrates the memory allocation for Rec_fact at this point in the calculation. There are now two separate blocks of memory, one for the current
case $n = 3$ and one for the previous case $n = 4$ which is not yet done and
remains in memory. Notice that each block has its own storage locations
for the input variable n and the local variable f. In the computer's memory, these two blocks reside in an internal data structure known as a *stack*.

Briefly, a stack is a data structure for which data (or blocks of data) can only be inserted or removed from the top of the stack[1]. In this case, the top of the stack ($n = 3$) contains the active version of *Rec_fact*, and lower levels of the stack contain previous versions of *Rec_fact*, which have been invoked but are not yet done. For $n = 3$, the local variable f has not been assigned, and "next statement executed" refers to line 4 in the previous version *Rec_fact*(4) which invoked *Rec_fact*(3).

Now, to compute *Rec_fact*(3), we need the value of *Rec_fact*(2), which means we again invoke *Rec_fact* and assign yet another block of memory to the procedure. To complete the calculation, we continue invoking the procedure for successively smaller integer values until the termination condition $n = 0$ is reached. The memory allocation stack at this point is shown in Figure 5.2(c). Observe that the currently active version ($n = 0$) is at the top of the stack, and the other levels of the stack represent the previous procedure calls that led to this place in the calculation. At this point, the variable f (for the $n = 0$ case) is assigned the value 1 (with lines 1, 2), and this value is returned as the value of *Rec_fact*(0). Once this is done, the block of memory allocated for *Rec_fact*(0) is no longer needed and is removed from the top of the stack. Control is now transferred back to line 4 in *Rec_fact*(1) which performs the multiplication and assignment:

$$f := 1 * Rec\_fact(0) \rightarrow 1 * 1 \rightarrow 1 \quad \text{(calculation in } Rec\_fact(1)).$$

This value is returned to *Rec_fact*(2) which invoked *Rec_fact*(1), and the memory allocated for *Rec_fact*(1) is removed from the top of the stack. The recursive process continues to unwind in this fashion, performing the multiplication and assignment in line 4 for the different versions of *Rec_fact*:

$$f := 2 * Rec\_fact(1) \rightarrow 2 * 1 \rightarrow 2 \quad \text{(calculation in } Rec\_fact(2)),$$
$$f := 3 * Rec\_fact(2) \rightarrow 3 * 2 \rightarrow 6 \quad \text{(calculation in } Rec\_fact(3)),$$
$$f := 4 * Rec\_fact(3) \rightarrow 4 * 6 \rightarrow 24 \quad \text{(calculation in } Rec\_fact(4)).$$

In each case, once an expression has been returned by *Rec_fact*($n - 1$) to the calling procedure *Rec_fact*(n) (or the interactive mode), the block of memory associated with *Rec_fact*($n - 1$) is removed from the top of the stack. After the last calculation, the expression 24 is returned as the value of *Rec_fact*(4).

The *Rec_fact* procedure is presented to illustrate simply what is meant by a recursive procedure and to show how it is evaluated by a CAS. In practice, the recursive procedure for $n!$ is less efficient in terms of computer time and memory than a non-recursive iterative procedure.

---

[1] A useful metaphor for a stack data structure is a stack of food trays. For safety's sake, we always remove a tray from the top of the stack and add a tray to the stack by placing it on the top.

Infinite Recursive Loops

A call to $Rec\_fact(n)$ terminates as long as n is a non-negative integer. However, if n is a negative integer (or any expression that does not evaluate to a positive integer), the termination condition in line 1 is never satisfied, and so the process does not terminate. For example, when $n = -1$, we obtain the infinite sequence of procedure calls:

$$Rec\_fact(-1), \ Rec\_fact(-2), \ Rec\_fact(-3), \ldots.$$

Since this problem is similar to the infinite loops that can arise with iteration structures, it is called an *infinite recursive loop*.

Exercises

1. Let n be a positive integer. The harmonic number $H(n)$ is defined by the sum:
$$H(n) = 1 + 1/2 + \cdots + 1/n.$$
Give a recursive procedure for $H(n)$. The procedure should not use a **for** structure or a **while** structure.

2. The Fibonacci number sequence $f_0, f_1, f_2, \ldots$ is defined using the recursive definition:
$$f_n = \begin{cases} 1, & \text{when } n = 0 \text{ or } n = 1, \\ f_{n-1} + f_{n-2}, & \text{when } n > 1. \end{cases} \tag{5.3}$$

   (a) Compute f_4.

   (b) Here is a recursive MPL procedure that computes the Fibonacci numbers:

```
        Procedure  Fibonacci(n);
        Input
            n : non-negative integer;
        Output
            fn;
        Local Variables
            f, g, r;
        Begin
    1       if n = 0 or  n = 1 then
    2           r = 1
    3       else
    4           f := Fibonacci(n - 1);
    5           g := Fibonacci(n - 2);
    6           r := f + g;
    7       Return(r)
        End
```

Trace the flow of the *Fibonacci* procedure for $n = 4$ showing all changes in the storage allocation stack during the course of the computation.

(c) Give a non-recursive procedure that uses iteration to compute f_n.

(d) The Fibonacci computation is not a particularly good use of recursion since the non-recursive approach requires fewer additions than the recursive approach. Explain why this is so.

3. Let S be a non-empty set that contains n expressions, and for $0 \leq k \leq n$ let $C(n, k)$ be the number of distinct subsets of size k of S. We can obtain $C(n, k)$ using the familiar combination formula

$$C(n, k) = \frac{n!}{k!(n - k)!}.$$

$C(n, k)$ can also be obtained recursively using the recurrence relation

$$C(n, k) = \begin{cases} 1, & \text{if } k = 0 \text{ or } k = n, \\ C(n - 1, k - 1) + C(n - 1, k), & \text{otherwise.} \end{cases} \quad (5.4)$$

Give a procedure for $C(n, k)$ that is based on Expression (5.4). Do not use the factorial operation in this procedure.

5.2 Recursive Procedures

In this section we give a number of examples that illustrate the possibilities and limitations of recursion as an algorithmic approach for computer algebra.

The *Complete_sub_expressions* Operator

In this example we describe a procedure that obtains the set of complete sub-expressions of an expression u. Since the solution of this problem involves a systematic traversal of the expression tree for u, a recursive procedure is the natural choice.

An MPL procedure that performs this operation is given in Figure 5.3. Lines 1-2, which apply to atomic expressions, provide the termination condition for the recursion. For compound expressions, the statements in lines 4-7 obtain the set of sub-expressions by forming the set union of $\{u\}$ and the sets of sub-expressions of the operands of u.

Let's see how the procedure works for $u = a * (x + 1) + 3 * \cos(y)$, which is represented by the expression tree in Figure 5.4. The flow of the computation in response to the statement

$$Complete\_sub\_expressions(a * (x + 1) + 3 * \cos(y)) \quad (5.5)$$

Procedure *Complete_sub_expressions(u)*;
Input
    u : a mathematical expression;
Output
    the set of complete sub-expressions of *u*;
Local Variables
    s, i;
Begin
1   **if** *Kind(u)* ∈ {**integer, symbol, real**} **then**
2   *Return({u})*
3   **else**
4   *s* := {*u*};
5   **for** *i* := 1 **to** *Number_of_operands(u)* **do**
6   *s* := *s* ∪ *Complete_sub_expressions(Operand(u, i))*;
7   *Return(s)*
End

Figure 5.3. An MPL procedure that finds the set of complete sub-expressions of
u. (Implementation: Maple (txt), Mathematica (txt), MuPAD (txt).)

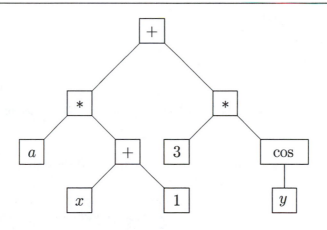

Figure 5.4. An expression tree for $a * (x + 1) + 3 * \cos(y)$.

is shown in Figure 5.5. The arrows that point downward on solid lines
represent a recursive call to a procedure, and those that point upward on
dashed lines represent a return to the calling procedure. The expressions at
the nodes represent the input expression *u* on various calls of the procedure,

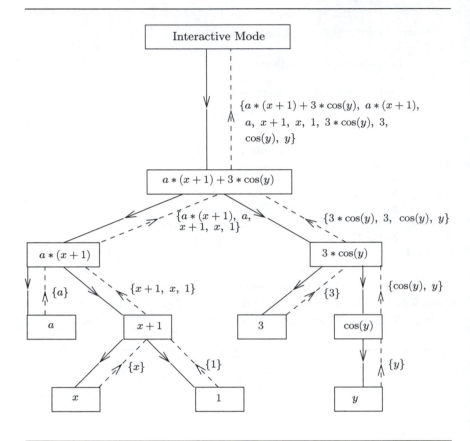

Figure 5.5. The sequence of recursive calls that obtains the set of complete sub-expressions of $a * (x + 1) + 3 * \cos(y)$.

and the sets of expressions to the right of the dashed lines above the nodes represent the output of that call.

By tracing the path along the solid and dashed lines, we observe the entire path of the computation. For example, to evaluate Expression (5.5), the procedure must first evaluate

$$Complete\_sub\_expressions(a * (x + 1)), \qquad (5.6)$$

$$Complete\_sub\_expressions(3 * \cos(y)). \qquad (5.7)$$

Observe that the entire computation associated with (5.6) is done before (5.7) is invoked, and to obtain (5.6), the procedure must evaluate

$$Complete\_sub\_expressions(a),$$
$$Complete\_sub\_expressions(x+1).$$

Continuing in this fashion, we systematically build up the set of sub-expressions of $a * (x + 1) + 3 * \cos(y)$ to obtain

$$\{a * (x + 1) + 3 * \cos(y), a * (x + 1), a, x + 1, x, 1, 3 * \cos(y), 3, \cos(y), y\}.$$

The *Free_of* Operator

The procedure for the *Free_of*(u, t) operator (see Definition 3.28, page 110) is another example that utilizes the recursive tree structure of an expression. Recall that the operator returns **false** when t is syntactically equal to a complete sub-expression of u, and otherwise returns **true**.

An MPL procedure for the *Free_of* operator is given in Figure 5.6. Lines 1 and 3 serve as terminating conditions for the procedure. If the condition in line 3 is **true**, the procedure returns **true** because the condition in line 1 is **false** and u does not have any operands. The loop (lines 7-10) applies

---

Procedure *Free_of*(u, t);
Input
    u, t : mathematical expressions;
Output
    true or false;
Local Variables
    i;
Begin
1   if $u = t$ then
2   *Return*(**false**)
3   elseif *Kind*$(u) \in \{$**symbol, integer, real**$\}$ then
4   *Return*(**true**)
5   else
6        $i := 1$;
7   while $i \leq$ *Number_of_operands*(u) do
8   if not *Free_of Operand*$(u, i), t)$ then
9   *Return*(**false**);
10           $i := i + 1$;
11   *Return*(**true**)
End

Figure 5.6. An MPL procedure for the *Free_of* operator. (Implementation: Maple (txt), Mathematica (txt), MuPAD (txt).)

the procedure recursively to each operand of u. Notice when a recursive call on some operand returns **false**, there is no need to check the remaining operands and so the value **false** is returned immediately. If all operands of u are free of t, the procedure returns **true** (line 11).

In the current form, the *Free_of* operator cannot determine if an expression is free of an algebraic operator or function name. A modification of the procedure that handles these cases is described in Exercise 1(b).

A useful extension of the *Free_of* operation is to check if u is free of each expression in a set (or list) S of expressions. The procedure for *Set_free_of*(u, S) that performs this operation is a simple modification of the one for *Free_of*(u, t). The details of this extension are left to the reader (Exercise 1(c)).

Pattern Matching, the *Linear_form* Operator

Many operations in mathematics depend on recognizing that an expression has a particular form. In this example we describe $Linear\_form(u, x)$, a simple pattern-matching procedure that checks if an algebraic expression u has the form $a\,x + b$, where the expressions a and b are free of x. When this is so, the procedure returns the list $[a, b]$, and otherwise returns the global symbol **Fail**. We interpret this form in a broad sense to include more involved sums (e.g., $a\,x + 2\,x + b + 3$) as well as expressions that are not sums (e.g., 3, x, $2\,x$, x/a).

An MPL procedure for this operation is shown in Figure 5.7. Lines 1-4 handle two simple cases that have the required form. Lines 5-11 check the form of a product, where lines 8-9 check if the symbol x is an operand of the product. In lines 12-21, recursion is used to check if the operands of a sum have the proper form. To do this, we apply the operator to the first operand of the sum (line 13) and the remaining operands (line 17), and then combine the results (line 21). If some operand of the sum does not have the proper form, the symbol **Fail** is returned (lines 15, 19). Lines 22-25 handle other expression types (e.g., powers, function forms, factorials), which only have the proper form when they are free of x.

There are two places in this procedure where recursion is used, lines 13 and 17. We can eliminate this recursion by using an iteration structure to check the operands of a sum and by repeating the statements for the tests in lines 1-11 and 22-25. Although recursion can be eliminated here, it is used as a matter of convenience to obtain a shorter procedure.

Pattern-matching procedures are given for quadratic polynomials in Exercise 8, and for more general polynomials in Chapter 6.

Procedure   *Linear_form*(u, x);
Input
    u : an algebraic expression;
    x : a symbol;
Output
    the list [a, b], where a and b are algebraic expressions, or the global
    symbol **Fail**;
Local Variables
    f, r;
Begin
1   **if** u = x **then**
2   *Return*([1, 0])
3   **elseif** *Kind*(u) ∈ {**symbol, integer, fraction**} **then**
4   *Return*([0, u])
5   **elseif** *Kind*(u) = " * " **then**
6   **if** *Free_of*(u, x) **then**
7   *Return*([0, u])
8   **elseif** *Free_of*(u/x, x) **then**
9   *Return*([u/x, 0])
10   **else**
11   *Return*(**Fail**)
12   **elseif** *Kind*(u) = " + " **then**
13   f := *Linear_form*(*Operand*(u, 1), x);
14   **if** f = **Fail then**
15   *Return*(**Fail**)
16   **else**
17   r := *Linear_form*(u − *Operand*(u, 1), x);
18   **if** r = **Fail then**
19   *Return*(**Fail**)
20   **else**
21   *Return*([*Operand*(f, 1) + *Operand*(r, 1),
                         Operand(f, 2) + *Operand*(r, 2)])
22   **elseif** *Free_of*(u, x) **then**
23   *Return*([0, u])
24   **else**
25   *Return*(**Fail**)
End

Figure 5.7. An MPL procedure that determines if u is a linear expression in x. (Implementation: Maple (txt), Mathematica (txt), MuPAD (txt).)

Transformation Rule Sequences, the *Derivative* Operator

In this example we describe an algorithm that computes the derivative of a function. Since the differentiation rules for sums, products, powers, and composite functions obtain the derivative of an expression in terms of the derivatives of its operands, the algorithm is recursive.

For this example, we describe the algorithm using a *transformation rule sequence* rather than an MPL procedure. The description is somewhat simpler in this format, and the transformation rules can be easily translated into an MPL procedure. In addition, some CAS languages have the capability to implement transformation rules directly as a program.

Let u be an algebraic expression and let x be a symbol. The operator $Derivative(u, x)$, which evaluates the derivative of u with respect to x, is defined by the following transformation rules:

DERIV-1. If $u = x$, then $Derivative(u, x) \to 1$.

DERIV-2. If $u = v^w$, then

$$Derivative(u, x) \to \qquad\qquad\qquad\qquad\qquad\qquad (5.8)$$
$$w * v^{w-1} * Derivative(v, x) + Derivative(w, x) * v^w * \ln(v).$$

This rule applies to expressions that are powers and accounts for expressions where either v or w may depend on x. (The rule is derived using logarithmic differentiation (Exercise 12).) Since the *Derivative* operator appears on the right side of the rule, DERIV-2 is recursive. When w is free of x, the rule reduces (with automatic simplification) to the familiar power rule

$$\frac{d(v^w)}{dx} = w \cdot v^{w-1} \frac{d(v)}{dx}.$$

DERIV-3. Suppose u is a sum and let $v = Operand(u, 1)$ and $w = u - v$. Then

$$Derivative(u, x) \to Derivative(v, x) + Derivative(w, x).$$

DERIV-4. Suppose u is a product and let $v = Operand(u, 1)$ and $w = u/v$. Then

$$Derivative(u, x) \to Derivative(v, x) * w + v * Derivative(w, x).$$

Rules DERIV-3 and DERIV-4 are the sum and product differentiation rules. Again, the rules are recursive because the right side of each rule refers to the *Derivative* operator. Notice that we obtain the derivative of a sum by differentiating both the first operand and the remaining part of the sum, which is obtained by subtracting the first operand from u with automatic simplification. A similar approach is used for a product.

A typical rule for a known function looks like the following:

DERIV-5. If $u = \sin(v)$, then $Derivative(u, x) \to \cos(v) * Derivative(v, x)$.

Again, the chain rule implies the rule is recursive.

DERIV-6. If $Free\_of(u, x) = \textbf{true}$, then $Derivative(u, x) \to 0$.

This rule applies to integers, fractions, symbols, and compound expressions (such as $f(a)$ or $n!$) that are free of the differentiation variable x. Notice that powers, sums, and products are not checked by this rule because they are handled by one of the earlier rules DERIV-2, DERIV-3, or DERIV-4. For example, if b and e are symbols ($\neq x$), then

$$Derivative(b^e, x) \to 0$$

is obtained by first applying DERIV-2, which applies DERIV-6 (recursively) to both b and e.

We have placed DERIV-6 at this point in the rule sequence to avoid redundant calls on the $Free\_of$ operator. The reason for this has to do with the recursive nature of $Free\_of$. If DERIV-6 were at the beginning of the rule sequence, then to compute the derivative (with respect to x) of

$$u = (1 + a)^2 + x^2,$$

the algorithm would first check if u were free of x, which involves the comparison of each complete sub-expression of u to x until the symbol x is found. Since this step would return **false**, we would next apply the sum rule which obtains the derivative in terms of the derivatives of the two operands $(1 + a)^2$ and x^2. To find the derivative of $(1 + a)^2$, we would check (for the second time) if this expression were free of x. By placing the $Free\_of$ operation later in the rule sequence, we avoid this redundant calculation.

The final transformation rule applies to any expression that is not covered by the earlier rules:

DERIV-7. $Derivative(u, x) \to "Derivative"(u, x)$.

In other words, if none of the earlier rules apply to u, the expression is returned in the unevaluated form $Derivative(u, x)$. The $Derivative$ operator on the right is quoted to prevent a recursive evaluation of the operator because, without the quotes, the transformation leads to an infinite sequence

of recursions. By including this rule, we obtain a representation for the
derivative of expressions that include undefined functions such as

$$Derivative(f(x) * g(x), x) \quad \rightarrow \quad Derivative(f(x), x) * g(x) \qquad (5.9)$$
$$+ f(x) * Derivative(g(x), x),$$

where the derivatives of $f(x)$ and $g(x)$ remain in unevaluated form. (See
Exercise 13(c) for an extension of this situation.)

Notice that the differentiation quotient rule is not included in our rule
sequence because we assume that automatic simplification transforms quotients to products or powers. In some instances, however, the quotient rule
returns the derivative of a quotient in a more useful form. Since it is not
difficult to check when a product is a quotient, this is a useful extension of
the algorithm (Exercise 13(b)).

The DERIV rules are an example of a transformation rule sequence.
When describing an algorithm in this way, we assume that a rule is checked
only when all earlier rules do not apply. This approach simplifies the presentation because conditions that are handled by earlier rules need not be
repeated (in a negative sense) in a later rule.

It is a simple matter to express the DERIV rule sequence as an MPL
procedure. We leave the details of the procedure to the reader (Exercise 13).

Rule-Based Programming

Some CAS languages have the capability to implement a transformation
rule sequence directly.

Mathematica. Figure 5.8 shows an implementation of the DERIV rules in
the Mathematica pattern matching language. Since `Derivative` is a predefined operator in this system, we have used the name `Deriv` instead.

```
Deriv[x_, x_ ] := 1;
Deriv[ v_^w_, x_] := w*v^(w-1)*Deriv[v,x] + Deriv[w,x]*v^w*Log[v];
Deriv[ u_ + v_, x_ ] :=  Deriv[u,x] + Deriv[v,x];
Deriv[ u_ * v_, x_ ] :=  Deriv[u,x]*v + Deriv[v,x]*u;
Deriv[Sin[u_], x_ ] := Cos[u]*Deriv[u,x];
Deriv[u_,x_] := 0 /;  FreeQ[u,x] === True;
```

Figure 5.8. A rule-based program for the *Derivative* operator in the Mathematica
pattern matching language. Since `Derivative` is a predefined operator in the
Mathematica language, we have used the name `Deriv` instead. (Implementation:
Mathematica (nb).)

In Mathematica, an underscore character (_) after a variable name means the variable can stand for an arbitrary expression. The symbol /; (in the last line) stands for the word "whenever," and so the free of condition following this symbol must hold for the rule to apply. Mathematica keeps re-applying the rules to an expression until changes do not occur. For this reason, even though the sum and product rules are listed with only two operands, the operator can differentiate sums or products with more than two operands as well. Notice that we have omitted the last rule DERIV-7 because if u does not satisfy one of the input patterns, Mathematica returns the operator in the unevaluated form `Deriv[u,x]`.

Once the transformation rules have been entered in a Mathematica session, they are applied during evaluation whenever the `Deriv` operator appears in an expression. In Mathematica, the execution order for rules does not depend on the order in which they are listed. Rather, the system applies more specific rules before it applies more general rules. For this example, however, the rule that involves the `FreeQ` operator is checked after the other rules.

Maple. Figure 5.9 shows an implementation of the DERIV rules in the Maple pattern matching language. Notice that each symbol (x, u, v, and w) is followed by two colons (::) and one of the designations

name, algebraic, nonunit(algebraic)

which defines the class of expressions that can replace the variable. The form `nonunit(algebraic)` is included so that an expression is not matched

---

```
define(Derivative,
Derivative(x::name,x::name)=1,
Derivative(v::nonunit(algebraic)^w::nonunit(algebraic),x::name)
        =w*v^(w-1)*Derivative(v,x)+Derivative(w,x)*v^w*ln(v),
Derivative(u::nonunit(algebraic)+v::nonunit(algebraic),x::name)
        =Derivative(u,x)+Derivative(v,x),
Derivative(u::nonunit(algebraic)*v::nonunit(algebraic),x::name)
        =Derivative(u,x)*v+Derivative(v,x)*u,
Derivative(sin(u::algebraic),x::name)=cos(u)*Derivative(u,x),
conditional(Derivative(u::algebraic,x::name)
        =0,_type(u,freeof(x)))
);
```

Figure 5.9. A rule-based program for the *Derivative* operator in the Maple pattern-matching language. (Implementation: Maple (mws).)

by an inappropriate rule. For example, this form is included in the product rule so that the Maple's pattern matching algorithm does not consider the expression $\sin(x)$ to be a product $1 * \sin(x)$. (Without this designation, the execution of `Derivative(sin(x), x)` results in an infinite recursive loop.) The `nonunit` designation also permits sums and products in rules 3 and 4 to have more than two operands. The `conditional` statement in the last rule implements the *Free_of* test in DERIV-6. Notice that DERIV-7 is not needed because when u does not match any of the rules, `Derivative(u, x)` is returned in unevaluated form.

In Maple, the transformation rules are checked in the order they are listed, and once the rules have been entered in a session, the system creates a recursive procedure with the name `Derivative`.

Rule-based programming usually gives smaller programs because much of the program logic is handled by the CAS's pattern matching program. On the other hand, because program logic is handled by the system, we give up some control of the process. In addition, the approach requires a good understanding of the workings (and limitations) of the pattern matching program, and, in some cases, it can be difficult (or even impossible) to express a transformation in the required form.

The *Trig_substitute* Operator

Let u be an algebraic expression. The operator $Trig\_substitute(u)$ forms a new expression, with all instances of the functions tan, cot, sec, and csc in u replaced by the equivalent representations in terms of sin and cos.

The operator utilizes the four transformation rules:

TRIGSUB-1. $\tan(v) \rightarrow \dfrac{\sin(v)}{\cos(v)}$.

TRIGSUB-2. $\cot(v) \rightarrow \dfrac{\cos(v)}{\sin(v)}$.

TRIGSUB-3. $\sec(v) \rightarrow \dfrac{1}{\cos(v)}$.

TRIGSUB-4. $\csc(v) \rightarrow \dfrac{1}{\sin(v)}$.

The easiest way to obtain these transformations is with the rule-based operations that are available in some CAS languages. It is instructive, however, to obtain the transformations with MPL procedures. We describe two approaches, one based on the *Construct* operator described in Section 3.2 and the other based on the *Map* operator described below.

Procedure *Trig_substitute*(*u*);
Input
    u : an algebraic expression;
Output
    a new expression, with all instances of the functions
    tan, cot, sec, and csc replaced by the representations
    using sin and cos;
Local Variables
    s, i, L;
Begin

```
1     if  Kind(u) ∈ {integer, fraction, symbol} then
2         Return(u)
3     else
4         L := [ ];
5         for  i := 1 to  Number_of_operands(u) do
6             L := Join(L, [Trig_substitute(Operand(u, i))]);
7         if  Kind(u) ∈ {tan, cot, sec, csc} then
8             s := Operand(L, 1);
9             if  Kind(u) = tan then
10                Return(sin(s)/ cos(s));
11            if  Kind(u) = cot then
12                Return(cos(s)/ sin(s));
13            if  Kind(u) = sec then
14                Return(1/ cos(s));
15            if  Kind(u) = csc then
16                Return(1/ sin(s))
17        else
18            Return(Construct(Kind(u), L))
      End
```

Figure 5.10. An MPL procedure for *Trig_substitute* that uses the *Construct* operator. (Implementation: Maple (txt), Mathematica (txt), MuPAD (txt).)

A *Trig_substitute* procedure[2] that uses the *Construct* operator is given in Figure 5.10. Lines 1-2 provide a termination condition for the recursion. In lines 4-6, we construct a list *L* that contains the expressions obtained

---

[2] This procedure will not work in the Mathematica system using this system's trigonometric functions (`Sin[x]`, `Cos[x]`, `Tan[x]`, etc.) because the automatic simplification rules cancel the operations of the procedure. For example, in this system, automatic simplification obtains the inverse transformation replacing `Sin[x]/Cos[x]` with `Tan[x]`. To implement the procedure, it is necessary to override the automatic simplification rules by using different names for these functions. One possibility is to use function names that begin with lower case characters. (Implementation: Mathematica (nb).)

by applying *Trig_substitute* to each operand of u. Lines 7-16 apply the TRIGSUB transformations where, for these cases, the operand list L has only one operand. For all other compound expressions, we construct (line 18) a new expression using the same main operator as u and the operands of L.

For expressions whose main operator is an algebraic operator ($+$, $*$, $\wedge$, or $!$), it is not necessary to use the *Construct* operator and the iteration structure in lines 5-6. For example, for sums we obtain the same result by returning the expression

$$Trig\_substitute(Operand(u, 1)) + Trig\_substitute(u - Operand(u, 1)).$$

However, we have used the *Construct* operator because each operator requires its own statement similar to this one. In addition, we must use *Construct* for function forms (such as $f(\tan(x), \sec(x) + 1)$) that can have an arbitrary number of operands but don't satisfy an algebraic relation.

The *Map* Operator

A basic operation in the *Trig_substitute* procedure is the creation of a new expression with the same main operator as u and operands that are obtained by recursively applying the procedure to each operand of u. Since this operation occurs frequently in computer algebra, it is useful to have an MPL primitive operator that performs the operation. The *Map* operator serves this purpose.

Definition 5.1. *Let u be a compound expression with*

$$n = Number\_of\_operands(u),$$

and let $F(x)$ and $G(x, y, \ldots, z)$ be operators. The Map operator has two forms:

$$Map(F, u), \tag{5.10}$$

$$Map(G, u, y, \ldots, z). \tag{5.11}$$

The statement $Map(F, u)$ obtains the new expression with main operator $Kind(u)$ and operands

$$F(Operand(u, 1)), \ F(Operand(u, 2)), \ldots, F(Operand(u, n)).$$

The statement $Map(G, u, y, \ldots, z)$ obtains the new expression with main operator $Kind(u)$ and operands

$$G(Operand(u, 1), y, \ldots, z), \ G(Operand(u, 2), y, \ldots, z), \ldots,$$
$$G(Operand(u, n), y, \ldots, z).$$

MPL	Maple	Mathematica	MuPAD
$Map(F,$ $a+b)$	Map(F,a+b)	Map[F,a+b]	Map(a+b,F)
$Map(G,$ $a+b,d,e)$	Map(G,a+b,d,e)	Map[G[#,d,e]&,a+b]	Map(a+b,G,d,e)

Figure 5.11. The syntax of *Map* operators in Maple, MuPAD, and Mathematica. (Implementation: Maple (mws), Mathematica (nb), MuPAD (mnb).)

If u is not a compound expression, the Map operator returns the global symbol **Undefined**.

Example 5.2. For the operator

$$F(x) \overset{\text{function}}{:=} x^2,$$

we have

$$Map(F, a + b) \rightarrow a^2 + b^2.$$

For the operator

$$G(x, y, z) \overset{\text{function}}{:=} x^2 + y^3 + z^4,$$

we have

$$
\begin{aligned}
Map(G, a + b, c, d) \quad &\rightarrow \quad G(a, c, d) + G(b, c, d) \\
&= \quad \left(a^2 + c^3 + d^4\right) + \left(b^2 + c^3 + d^4\right) \\
&= \quad a^2 + b^2 + 2\,c^3 + 2\,d^4. \qquad \qquad \square
\end{aligned}
$$

Most CAS languages have some form of the *Map* operator (Figure 5.11). A procedure[3] for trigonometric substitution that uses the *Map* operator is given in Figure 5.12.

Computation of Legendre Polynomials

This example provides another simple example of the mechanics of recursion and reveals one of its limitations.

The Legendre polynomials are the sequence of polynomials $p_n(x), n = 0, 1, 2, \ldots$ that are defined by the relations

$$
\begin{aligned}
p_0(x) &= 1, & (5.12) \\
p_1(x) &= x, & (5.13) \\
p_n(x) &= \frac{1}{n}((2n - 1)\,x\,p_{n-1}(x) - (n - 1)\,p_{n-2}(x)), n \geq 2. & (5.14)
\end{aligned}
$$

[3]Mathematica users see footnote 2 on page 187.

   Procedure *Trig_substitute_map*(*u*);
   Input
     u : an algebraic expression;
   Output
     a new expressions where all instances of the functions
     tan, cot, sec, and csc are replaced by the representations
     using sin and cos;
   Local Variables
     U;
   Begin

```
1     if  Kind(u) ∈ {integer, fraction, symbol} then
2         Return(u)
3     else
4         U := Map( Trig_substitute_map, u);
5         if  Kind(U) = tan then
6             Return(sin( Operand(U, 1))/ cos( Operand(U, 1)));
7         if  Kind(U) = cot then
8             Return(cos( Operand(U, 1))/ sin( Operand(U, 1)));
9         if  Kind(U) = sec then
10            Return(1/ cos( Operand(U, 1)));
11        if  Kind(U) = csc then
12            Return(1/ sin( Operand(U, 1)))
13        else
14            Return(U)
```

   End

Figure 5.12. An MPL procedure for trigonometric substitution that uses the *Map* operator. (Implementation: Maple (txt), Mathematica (txt), MuPAD (txt).)

The polynomials are named in honor of the French mathematician Adrien-Marie Legendre (1752-1833), who first used them in 1785 to study the gravitational attraction of solids of revolution. Today they have applications in numerical integration, the solution of differential equations, and engineering.

    The expression for $p_n(x)$ is called a *recurrence relation* because for $n \geq 2$, $p_n(x)$ is defined in terms of the lower order polynomials $p_{n-1}(x)$ and $p_{n-2}(x)$. The polynomials p_0 and p_1 serve as termination conditions for the recursion. Using this definition, each succeeding polynomial ($n \geq 2$) is computed as follows:

$$p_2(x) \quad = \quad \frac{1}{2}((2(2) - 1)\, x\, p_1(x) - (2 - 1)\, p_0(x)) = \frac{3}{2}x^2 - \frac{1}{2},$$

$$p_3(x) \quad = \quad \frac{1}{3}((2(3) - 1)\, x\, p_2(x) - (3 - 1)\, p_1(x) = \frac{5}{2}x^3 - \frac{3}{2}x,$$

$$\vdots$$

etc.

A recursive procedure for $p_n(x)$ is given in Figure 5.13. Lines 6-7 contain recursive calls of the procedure, and line 8 contains an *Algebraic_expand* operator so that the polynomial is returned in expanded form.

Unfortunately, the *Legendre* procedure performs an excessive amount of redundant calculation that makes it unsuitable for large values of n. A trace of the recursive calls for *Legendre*$(4, x)$ indicates why this is so (see Figure 5.14). To compute *Legendre*$(4, x)$, the procedure must compute recursively *Legendre*$(3, x)$ and *Legendre*$(2, x)$. Observe that all recursive calculations for *Legendre*$(3, x)$ are done before any of the calculations for this version of *Legendre*$(2, x)$. In addition, to compute *Legendre*$(3, x)$, the procedure must compute another version of *Legendre*$(2, x)$ and *Legendre*$(1, x)$. The computation continues in this fashion until it encounters one of the terminating conditions $n = 0$ or $n = 1$. The cause of the redundant calculations is apparent from the sequence of procedure calls shown in Fig-

---

Procedure *Legendre*(n, x);
Input
    n : a non-negative integer;
    x: a symbol;
Output
    $p_n(x)$;
Local Variables
    f, g;
Begin
1   **if** $n = 0$ **then**
2   *Return*(1)
3   **elseif** $n = 1$ **then**
4   *Return*(x)
5   **else**
6        $f := $ *Legendre*$(n - 1, x)$;
7        $g := $ *Legendre*$(n - 2, x)$;
8   *Return*$(Algebraic\_expand((1/n) * ((2 * n - 1) * x * f - (n - 1) * g)))$
End

Figure 5.13. Computation of Legendre polynomials using recursion. (Implementation: Maple (txt), Mathematica (txt), MuPAD (txt).)

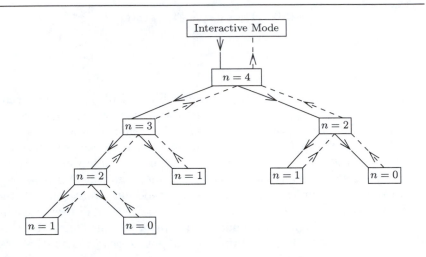

Figure 5.14. The sequence of recursive calls for *Legendre*(4, *x*). An arrow that points downward on a solid line represents a recursive call, and those that point upward on a dashed line represent a return to the calling procedure.

ure 5.14. Observe that there are two calls on *Legendre*(2, *x*), three calls on *Legendre*(1, *x*), and two calls on *Legendre*(0, *x*). In each instance, the procedure is not aware that the value is computed more than once. In general, the number of recursive calls on the procedure increases exponentially with *n*.

For this computation, it is a simple matter to avoid the redundant calculation by avoiding recursion altogether and using an iterative procedure. This is done by replacing lines 6-8 in Figure 5.13 by the iteration:

$$f := 1;$$
$$g := x;$$
for $i := 2$ **to** n **do**
$$\quad p := (1/i) * ((2 * i - 1) * x * g - (i - 1) * f);$$
$$\quad f := g;$$
$$\quad g := p;$$
$$Return(Algebraic\_expand(p));$$

After executing the loop, $p_n(x)$ is contained in the variable p, which is then expanded in the last line. Although the iterative version is based on the recurrence relation in Equation (5.14), it is not considered a recursive algorithm because it does not call itself directly or indirectly.

```
Legendre_remember := proc(x,n)
local f,g;
option remember;
if n = 0 then
  RETURN(1)
elif n = 1 then
  RETURN(x)
else
  f := Legendre_remember(x,n-1);
  g := Legendre_remember(x,n-2);
  RETURN(expand((1/n)*((2*n-1)*x*f - (n-1)*g)))
fi
end:
```

Figure 5.15. Computing Legendre polynomials with Maple using the option *remember*. The option is also available in Mathematica and MuPAD. (Implementation: Maple (txt), Mathematica (txt), MuPAD (txt).)

Since the potential for redundant calculations occurs frequently in computer algebra, it is useful to have a way to perform a calculation in a recursive manner that avoids the redundant calculation. For example, the Maple language has a feature called *option remember* that makes this possible (see Figure 5.15). When this option is declared within a Maple procedure, the system keeps a table of input/output expressions for all calls on the procedure. When the procedure is invoked, a check is made to see if the current input is identical to the input of a previous procedure call. When this is so, the output is returned from the value in the table. If the input expression is not in the table, the output is calculated in the usual way and the new input/output pair is stored in the table. Both Mathematica and MuPAD also have *remember* options for procedures.

For the computation of Legendre polynomials, the *remember* option dramatically reduces the redundant calculation. In many situations, however, redundant calculations can be eliminated by either avoiding recursion or by modifying the algorithm. For example, we avoided redundant calculations with the *Derivative* operator by placing the *Free_of* operation at the end of the transformation sequence (see page 183). For this reason, the *remember* feature is not used by any of the algorithms this book.

Recursive Chains

The procedures described so far are recursive because each procedure is defined directly in terms of another version of the same procedure. Recursion may also come about indirectly. For example, suppose a procedure u_1 does not call on itself directly, but calls on another procedure u_2 which then calls on u_1. In this case, u_1 is considered recursive because it calls on itself indirectly through the intervening procedure u_2. An example of a recursive chain is given in the case study in Section 5.3.

Exercises

1. (a) Trace the flow of the computation in response to the statement

    $$Free\_of(a\,(x+1) + 3\,\cos(y), x).$$

    (b) Modify the $Free\_of(u, t)$ operator so that it returns **false** when u contains a target t that is an algebraic operator or function name. For example,

    $$Free\_of(f(x) + y, \, f) \;\;\rightarrow\;\; \textbf{false},$$
    $$Free\_of(y + z, \text{ " + "}) \;\;\rightarrow\;\; \textbf{false}.$$

    (c) Give a procedure $Set\_free\_of(u, S)$ that determines if u is free of all expressions in a set (or list) S. The Set_free_of operator is used in the $Monomial\_gpe$ procedure in Figure 6.5 on page 227.

2. Give a procedure
    $$Trig\_free\_of(u)$$
    that returns the symbol **true** if an algebraic expression u is free of trigonometric functions (sin, cos, tan, cot, sec, csc) and the symbol **false** otherwise.

3. Let u be a mathematical expression. Give a procedure $Symbols(u)$ that returns the set of symbols in u.

4. Give a procedure
    $$Contain\_parameters(u, x)$$
    that returns **true** if the algebraic expression u contains any symbols other than the symbol x and **false** otherwise.

5. Let u be a mathematical expression. Give a procedure

    $$Algebraic\_expression(u)$$

    that returns **true** if u is an algebraic expression and **false** if it is not algebraic. (See Definition 3.17 on page 93.)

6. Let u be a polynomial in x with rational number coefficients. An efficient way to evaluate a polynomial numerically is to rewrite the polynomial in a nested form by introducing extra parenthesis. For example,

    $$u = 2\,x^3 + 3\,x^2 + 4\,x + 6 = ((2\,x + 3)\,x + 4)\,x + 6.$$

In numerical methods texts, this method for evaluating a polynomial is called Horner's method (see Epperson [35]). Give a recursive procedure $Horner(u, x)$ that transforms a polynomial from the expanded form to the nested form.

7. A **numerical** expression u is one that is defined by the rule sequence:

NUM-1. u is an integer or a fraction.

NUM-2. u is one of the symbols π or e.

NUM-3. u is a compound expression with main operator $+$, $*$, $\wedge$, or a function name (sin, f, etc.) such that each operand of u is a numerical expression.

For example, the following are numerical expressions

$$2 + 2^{1/2}, \quad \sin(3), \quad f(3), \quad 2 \cdot \pi^{1/3}, \quad 3 + e.$$

Give a procedure $Numerical(u)$ that returns **true** if an algebraic expression u is a numerical expression and otherwise returns **false**.

8. Let u be an algebraic expression and let x be a symbol. Give a procedure $Quadratic\_form(u, x)$ that determines if u has the form $a\,x^2 + b\,x + c$ where a, b, and c are free of x. If u has the proper form return $[a, b, c]$, otherwise return **Fail**. Interpret this form in a broad sense to include more involved sums (e.g., $a\,x^2 + 2\,x^2 + b\,x + 3\,x + 4$) as well as expressions that are not sums (e.g., x^2, $2\,x$, x/a, $a\,b$, and 3). The point of this exercise is to implement the procedure in terms of primitive operators ($Kind$, $Operand$, etc.) and structure-based operators ($Free\_of$), and not in terms of polynomial operators ($Degree$, $Coefficient$).

9. Let u be an algebraic expression. Define the $tree\text{-}size$ of u as the number of symbols, integers, algebraic operators, and function names that occur in u. For example, the expression $(x + \sin(x) + 2) * x^3$ consists of $x, +, \sin, x, 2, *, x, \wedge$, and 3 and so has a tree-size of 9. Give a procedure $Tree\_size(u)$ that obtains the tree-size of u.

10. Give procedures for each of the following operators. In each case the procedures should be defined in terms of the primitive operators as was done in the text with the $Trig\_substitute$ and $Trig\_substitute\_map$ operators.

   (a) Let u be a mathematical expression and v an equation. Give a procedure for the operator $Substitute(u, v)$ that performs structural substitution. (See Definition 3.30 on page 111.)

   (b) Let u be a mathematical expression and L a list of equations. Give a procedure $Sequential\_substitute(u, L)$ that performs sequential substitution. (See Definition 3.31 on page 114.)

   (c) Let u be a mathematical expression and S a set of equations. Give a procedure $Concurrent\_substitute(u, S)$ that performs concurrent substitution. (See Definition 3.34 on page 115.)

11. (a) Let S be a set of mathematical expressions and let k be an integer with $0 \leq k \leq$ *Number_of_operands*(S). Give a procedure *Comb*(S, k) that returns the set of all k element subsets of S. For example, if $S = \{a, b, c, d\}$, then

$$Comb(S, 2) \rightarrow \{\{a, b\}, \{a, c\}, \{a, d\}, \{b, c\}, \{b, d\}, \{c, d\}\}.$$

The procedure can be defined by the following recursive transformation rule sequence.

   i. If $k =$ *Number_of_operands*(S), then *Comb*$(S, k) \rightarrow \{S\}$.
   ii. *Comb*$(S, 0) \rightarrow \{\emptyset\}$.
   iii. Let $x =$ *Operand*$(S, 1)$, $T = S \sim \{x\}$, and $D = Comb(T, k - 1)$. For $D = \{S_1, \ldots, S_n\}$, let $E = \{S_1 \cup \{x\}, \ldots, S_n \cup \{x\}\}$. Then

$$Comb(S, k) \rightarrow Comb(T, k) \cup E.$$

   (b) The *power set* of a set S is the set of all subsets of S. Give a procedure *Power_set*(S) that obtains the power set of a set S.

12. Derive the general differentiation power rule in DERIV-2. *Hint:* Let $y = v^w$ and take logs of both sides of the expression.

13. Let u be an algebraic expression, and let x be a symbol.

   (a) Give a procedure *Derivative*(u, x) that utilizes the DERIV rules described in this section.

   (b) Although quotients are represented as powers or products, it is possible to recognize when an expression is a quotient and apply the quotient rule instead of the product rule. Modify the *Derivative* procedure so that it recognizes when an expression is a quotient and, when this is so, applies the quotient rule.

   (c) Although the DERIV transformation rules allow for the differentiation of some expressions with undefined functions (see Statement (5.9)), the rules don't handle expressions with compositions of undefined functions such as $f(g(x))$ or $h(x, g(x))$ in an adequate way. For these expressions the *Derivative* operator is returned in unevaluated form instead of with a representation that utilizes the chain rule. Some computer algebra systems give representations of derivatives for these expressions that utilize the chain rule. Experiment with a CAS to see how derivatives of these expressions are handled, and modify the *Derivative* procedure to handle these derivatives.

   (d) The DERIV transformation rules provide for the differentiation of any algebraic expression including factorials. (According to the rules, *Derivative*$(x!, x)$ is now returned in unevaluated form.) Although $x!$ is defined only when x is a non-negative integer, there is a generalization of the factorial operation that involves the gamma function

$(\Gamma(x+1) = x!)$ where x is no longer restricted in this way[4]. With this generalization, we can define transformation rules for the differentiation of factorial expressions. Experiment with a CAS to see how the differentiation operator handles factorials, and modify the *Derivative* procedure to handle these expressions.

14. Let u be an equation with both sides of the equation algebraic expressions, x and y are symbols, and n is a non-negative integer. Give a recursive procedure

$$Implicit\_derivative(u, y, x, n)$$

that obtains the nth derivative of y with respect to x. If n= 0, return u. Use either the differentiation operator in a CAS or the *Derivative* operator in Exercise 13 to perform the differentiations. (If a CAS has the capability to perform implicit differentiation, do not use this capability.) Assume that y is represented in the equation in function notation $y(x)$ and that *Free_of*$(u, y(x))$ is **false**. Do not use an iteration structure in this procedure. For example,

$$Implicit\_derivative(x^2 + y(x)^2 = 1,\ y, x, 3) \rightarrow -3\,\frac{x\left(x^2 + y(x)^2\right)}{y(x)^5}.$$

15. Let u be an algebraic expression, and let x and y be symbols. Give a procedure

$$Derivative\_order(u, x, y)$$

that determines the maximum order of the derivatives of y with respect to x in u. In this exercise d(y,x) represents the first derivative and for an integer $n \geq 2$, $d(y, x, n)$ represents the derivative of order n. (We use this representation rather than a representation such as $Derivative(y(x), x)$ to conform with the presentation in Section 4.3.) In addition, the order of the symbol y is 0, and the order of expressions without a y is -1. To simplify matters, if u contains function forms with the name d that contain operands different from those in $d(y, x)$ and $d(y, x, n)$, return the global symbol **Undefined**. For example,

$$Derivative\_order(d(y, x, 2) + x\,d(y, x) + 4\,y,\ x,\ y) \quad \rightarrow \quad 2,$$
$$Derivative\_order(x + y,\ x,\ y) \quad \rightarrow \quad 1,$$
$$Derivative\_order(x,\ x,\ y) \quad \rightarrow \quad -1,$$
$$Derivative\_order(d(y^2, x),\ x,\ y) \quad \rightarrow \quad \textbf{Undefined},$$
$$Derivative\_order(d(y, b),\ x,\ y) \quad \rightarrow \quad \textbf{Undefined}.$$

Note in the last two examples, the symbol **Undefined** is returned because the operands of the function form d are inappropriate.

The *Derivative_order* operator is used in Exercise 14 on page 240.

---

[4]See Spanier and Oldham [92], Chapter 43 for a description of the gamma function and its derivative.

16. (a) Let S be a set of rational numbers. Give a procedure $Max(S)$ that returns the maximum value in S. If S is empty, return the global symbol **Undefined**.

(b) Suppose now that S is a finite set of algebraic expressions. Generalize the procedure Max so that it determines the maximum value of the expressions in S that can be compared. For the purposes of this exercise two expressions f and g are *comparable* if $f - g$ is an integer or fraction, and $f > g$ when $f - g > 0$ in automatic simplification. If all the expressions in S are pairwise comparable, then return the maximum expression. If two or more expressions cannot be compared, then return an unevaluated form of Max. For example,

$$Max(\{a, 2, 3\}) \;\rightarrow\; Max(\{a, 3\}),$$
$$Max(\{m, m + 1\}) \;\rightarrow\; m + 1,$$
$$Max(\{3, Max(\{2, x\}), \}) \;\rightarrow\; Max(3, x),$$
$$Max(\{-5, m, m + 1, 2, 3, \sqrt{2}\}) \;\rightarrow\; Max(\{3, m + 1, \sqrt{2}\}).$$

Note that in the last example 3 and $\sqrt{2}$ cannot be compared because $3 - \sqrt{2}$ is not an integer or fraction in automatic simplification.

This procedure returns a reasonable result as long as the input data is appropriate. For example, if some of the expressions in S are complex number expressions, then the input is not appropriate (e.g., $S = \{2, \sqrt{-1}\}$).

The Max operator is used in Exercise 17 below and Exercise 13, page 239.

17. Let u be an algebraic expression and x a symbol. Give a procedure

$$Max\_exponent(u, x)$$

that returns the largest exponent of x in u. If some exponents of x are not integers or fractions, return an unevaluated Max function as described in Exercise 16. For example,

$$Max\_exponent(x^2 + x^3, x) \;\rightarrow\; 3,$$
$$Max\_exponent(x + x^{-1}, x) \;\rightarrow\; 1,$$
$$Max\_exponent(\sin(x^2 + x^m, x) \;\rightarrow\; Max(\{2, m\}),$$
$$Max\_exponent(x^{(x^2)}, x) \;\rightarrow\; Max(\{2, x^2\}).$$

18. The absolute value function satisfies the following four properties:

(a) $|a \cdot b| \rightarrow |a| \cdot |b|$.

(b) For n an integer, $|a^n| \rightarrow |a|^n$.

(c) For $\imath = \sqrt{-1}$, $|\imath| \rightarrow 1$.

(d) If an expression has the form $a + \imath b$, where $a \neq 0$ and $b \neq 0$ are free of $\imath$, then $|a + b\imath| \rightarrow (a^2 + b^2)^{1/2}$.

Let u be an algebraic expression. Give a procedure $Absolute\_value(u)$ that obtains the absolute value of integers and fractions and applies the above rules when u is not an integer or fraction. If u is not an integer or fraction or one the above forms, return the unevaluated form $"Absolute\_value"(u)$. For example,

$$Absolute\_value(-1/2) \to 1/2,$$
$$Absolute\_value(-2\,x) \to 2\,Absolute\_value(x),$$
$$Absolute\_value(x+y) \to Absolute\_value(x+y),$$
$$Absolute\_value(x+2\imath) \to (x^2+4)^{1/2}.$$

This procedure returns a reasonable result as long as the input data is appropriate. For example, since the procedure does not perform an analysis of involved expressions with radicals, it may return an inappropriate result such as

$$Absolute\_value\left(\sqrt{1-\sqrt{2-\sqrt{5}}}+\imath\right) \to \sqrt{2-\sqrt{2-\sqrt{5}}},$$

which is a complex number.

5.3 Case Study: An Elementary Indefinite Integration Operator

In this case study we describe an algorithm that evaluates $\int f(x)\,dx$ for a limited class of functions encountered in elementary calculus. The algorithm utilizes the following:

1. an integration table,

2. the linear properties of the indefinite integral,

3. the "substitution" or "change of variable" method that is based on the inversion of the chain rule, and

4. both expanded and unexpanded forms of the integrand $f(x)$.

For example, the algorithm can evaluate the integrals

$$\int 5\,x\,\sin\left(x^2\right)\cos\left(x^2\right)\,dx, \qquad \int (\cos(x)+2)\,(\sin(x)+3)\,dx.$$

The Integration Table

The integration table includes the following standard elementary forms.

1. Expressions that are free of the integration variable x.

2. Powers x^n, where n is free of the integration variable x. Since most computer algebra systems return $\int x^{-1}\, dx = \ln(x)$ (rather than the more general form $\ln|x|$), we include this form in the table.

3. The functions $\exp(x)$ and $\ln(x)$ and the power b^x, where b is free of the integration variable x.

4. The trigonometric functions.

5. More involved expressions that occur as derivatives of the trigonometric functions or their inverses. For example, $\sec(x)\tan(x)$ appears in the table because it is the derivative of $\sec(x)$.

Linear Properties

When the integrand is a product $f(x) = c\,g(x)$ with c free of x, the algorithm applies the linear property

$$\int f\, dx = \int c g\, dx = c \int g\, dx \tag{5.15}$$

and then evaluates recursively $\int g\, dx$. In some cases when the substitution method is used to evaluate an integral, this step may seem counterproductive (see Equations (5.19)-(5.21) below). It is required, however, to match expressions in the integration table, and its application does not hinder the substitution method algorithm (see Example 5.4 below).

When f is a sum, the algorithm applies the linear property

$$\int f\, dx = \int \sum_{i=1}^{n} f_i\, dx = \sum_{i=1}^{n} \int f_i\, dx \tag{5.16}$$

and then evaluates recursively each $\int f_i\, dx$.

The Substitution Method

The substitution method is a basic technique for evaluating integrals that most readers are undoubtedly familiar with from the study of calculus. The method depends on the inversion of the chain rule

$$\int u(v(x))\, v'(x)\, dx = \int u(v)\, dv = U(v(x)), \tag{5.17}$$

where $U'(v) = u(v)$. It has the potential to obtain an anti-derivative whenever the integrand is a product of the form

$$f = u(v(x))\, v'(x). \tag{5.18}$$

Once such a representation is chosen, the success of the method depends on the evaluation of the new integral $\int u(v)\, dv$. For example, to evaluate

$$\int 2x \cos\left(x^2\right)\, dx, \tag{5.19}$$

let

$$v(x) = x^2, \qquad u(v) = \cos(v). \tag{5.20}$$

Since $v' = 2x$,

$$\int 2x \cos\left(x^2\right)\, dx = \int \cos(v)\, dv = \sin(v) = \sin\left(x^2\right). \tag{5.21}$$

Notice that we have omitted the arbitrary constant of integration as is done in most CAS software as well as the procedures in this section.

Although this example illustrates a general approach, the technique is more involved in practice. The difficulty involves deciding how to choose a substitution that eliminates the original integration variable x. In some instances it is possible to represent the integrand in the form (5.18) in a number of ways and in others it may not be possible at all. For our algorithm, we need a set of trial substitutions and a way to test if a substitution is appropriate. Figure 5.16 shows some typical substitutions used to evaluate integrals using this method.

These examples suggest four possible forms for the substitution $v(x)$.

1. *Function forms.* In

$$\int \frac{(x+1)\ln(\cos((x+1)^2))\sin((x+1)^2)}{\cos((x+1)^2)}\, dx, \tag{5.22}$$

the expressions

$$\ln(\cos((x+1)^2)), \quad \sin((x+1)^2), \quad \cos((x+1)^2)$$

are function forms.

2. *Arguments of function forms.* In (5.22), the expressions

$$\cos((x+1)^2), \quad (x+1)^2$$

are arguments of function forms.

Integral	Substitution
$\displaystyle \int \sin(x)\cos(x)\ dx = \frac{\sin^2(x)}{2}$	$v = \sin(x)$
$\displaystyle \int 2x\cos\left(x^2\right)\ dx = \sin\left(x^2\right)$	$v = x^2$
$\displaystyle \int 2x\left(x^2+4\right)^5\ dx = \left(x^2+4\right)^6/6$	$v = x^2 + 4$
$\displaystyle \int \cos(x)2^{\sin(x)}\ dx = \frac{2^{\sin(x)}}{\ln(2)}$	$v = \sin(x)$

Figure 5.16. Evaluation of integrals using the substitution method.

3. *Bases of powers.* In (5.22), the expressions

$$\cos((x+1)^2), \quad x+1$$

are bases of powers. The first expression is a base because the denominator of the integrand in (5.22) has the internal representation

$$(\cos((x+1)^2))^{-1}.$$

4. *Exponents of powers.* In $\cos(x)\,2^{\sin(x)}$, the expression $\sin(x)$ is an exponent of a power. In (5.22), -1 and 2 are also exponents, but don't give useful substitutions.

Using these substitution forms, the trial substitutions for the integrand in (5.22) are

$$\ln(\cos((x+1)^2)), \quad \cos((x+1)^2), \quad (x+1)^2,$$
$$x+1, \quad \sin((x+1)^2), \quad -1, \quad 2. \tag{5.23}$$

The first four expressions give substitutions that transform the integrand to the form in (5.18), while the last three do not. For example, the first substitution

$$v = \ln(\cos((x+1)^2)),$$

transforms the integral to a form that is easily evaluated

$$\int \frac{(x+1)\ \ln(\cos{(x+1)^2})\ \sin((x+1)^2)}{\cos((x+1)^2)}\ dx$$

$$= \ -(1/2)\int v\ dv = (-1/4)\,v^2 = (-1/4)\left(\ln(\cos((x+1)^2))\right)^2.$$

Substitutions using the next three expressions in (5.23) also lead to simpler integrals although each one requires at least one additional substitution for evaluation. For example, if the substitution is $v(x) = \cos((x+1)^2)$, then

$$\int \frac{(x+1)\ \ln(\cos((x+1)^2))\ \sin((x+1)^2)}{\cos((x+1)^2)}\ dx$$

$$= \ -(1/2)\int \frac{\ln(v)}{v}\ dv,$$

where the last integral is evaluated with another substitution $w = \ln(v)$.

A procedure for the substitution method must perform the following steps.

1. Form the set P of possible substitutions that contains the function forms, function arguments, and bases and exponents of powers in f.

2. Check each $v(x)$ in P to determine if it is an appropriate substitution. There are two expressions that may be in P, but which can be eliminated immediately. They are $v(x) = x$ which is really no substitution at all, and the expressions $v(x)$ that are free of x. If f has the factored form

$$f = u(v(x)) \cdot v'(x)$$

for some $v(x)$ in P, the new integrand $u(v)$ is obtained by eliminating the factor $v'(x)$ from f and substituting a symbol v for the expression $v(x)$. This operation is obtained by

$$u(v) = Substitute\left(\frac{f}{v'(x)},\ v(x) = v\right), \qquad (5.24)$$

where the division operation is obtained with automatic simplification. For the process to work, the substitution and division must eliminate the original integration variable x from the integrand. This condition is verified by checking that $u(v)$ is free of x. If this is so, we complete the integration by evaluating recursively $\int u(v)\ dv$, and by substituting $v(x)$ for v. Because of the division in Expression (5.24), the substitution method is also called the *derivative divides* method.

Example 5.3. Consider again

$$\int 2x \cos\left(x^2\right) \, dx.$$

The possible substitutions are

$$P = \{x, 2, x^2, \cos\left(x^2\right)\}. \tag{5.25}$$

Since the first two expressions x and 2 are not useful substitutions, the third one $v(x) = x^2$ is tried. In this case Expression (5.24) gives $u(v) = \cos(v)$, which is free of x, and so the anti-derivative is obtained with

$$Substitute \left(\int \cos(v) \, dv, \ v = x^2 \right) \to \sin\left(x^2\right).$$

On the other hand, with the fourth expression $v(x) = \cos\left(x^2\right)$ in P, Expression (5.24) gives

$$u(v) = \frac{-\cos(v)}{\sin\left(x^2\right)},$$

which is not free of x and so this substitution does not work. $\qquad\square$

Expanded versus Unexpanded Integrands

There are instances where expansion of the integrand is required for evaluation and others where expansion leads to more a difficult integration. For example, to evaluate

$$\int (x+1)\,(x+2) \, dx, \tag{5.26}$$

it is necessary to expand the integrand. On the other hand, while the unexpanded form

$$\int (2x+1) \cos\left(x^2+x\right) \, dx \tag{5.27}$$

is easily evaluated with the substitution $v(x) = x^2 + x$, by expanding and applying the linear property (5.16), we obtain

$$\int (2x+1)\cos\left(x^2+x\right) \, dx = \int 2x \cos\left(x^2+x\right) \, dx + \int \cos\left(x^2+x\right) \, dx,$$

where the two integrals on the right cannot be evaluated using the elementary functions encountered in calculus.

To handle both (5.26) and (5.27), the algorithm first tries to evaluate an integral without expanding f, and if it is not successful, tries again after expanding f.

The Integration Algorithm

The *Integral* procedure, which serves as a main procedure for the algorithm, is shown in Figure 5.17. The procedure returns either $\int f\ dx$ or the global symbol **Fail** if it is unable to evaluate the integral. It calls on three procedures (lines 1, 3, and 5) that also return either an evaluated integral or the symbol **Fail**. The statement at line 1 invokes the *Integral_table* procedure, which compares f to a number of standard forms and serves as a termination condition for the recursion. The procedure *Integral_table* is left to the reader (Exercise 3(a)).

If f is not in the table, then at lines 2-3 the procedure *Linear_properties* determines if either Equation (5.15) or Equation (5.16) can be applied, and, if so, applies the appropriate rule. This procedure is recursive because it calls on *Integral* to evaluate the new integrals produced by the linear properties. The procedure *Linear_properties* is left to the reader (Exercise 3(b)).

If this step fails, the *Substitution_method* procedure is applied at line 5. This step is recursive because this procedure also calls on *Integral*. If this step fails, the integrand is expanded (at line 7), and if this produces a new expression, the procedure *Integral* is applied recursively at line 9.

The *Substitution_method* procedure is shown in Figure 5.17. Notice that the procedure uses a global mathematical symbol v to avoid using a local variable that would be used without being assigned. At line 1, the *Trial_substitutions* procedure creates a set P of possible substitutions (Exercise 3(c)). In lines 4-10, we check each candidate g in P as a possible substitution. Once one is found, the loop terminates and the procedure returns the evaluated integral. The procedure is recursive because it calls on *Integral* at line 9, which allows another check of the integration table and further application of the linear properties, substitution method, and expansion, all of which may be needed (Exercise 1).

There are two ways that the *Substitution_method* procedure can fail to obtain the integral: first, when none of the possible substitutions in P works, and next, when the free-of test at line 8 succeeds but the *Integral* operator at line 9 is unable to evaluate the new integral. In either case, the symbol **Fail** is returned at line 11.

Example 5.4. Consider the evaluation of

$$\int 2\,x\,\cos\left(x^2\right)\ dx.$$

Figure 5.18 shows the sequence of procedure calls that indicates the path taken by the algorithm to evaluate the integral. (There are other procedure

Procedure *Integral*(f, x);
Input
    f : an algebraic expression;
    x : a symbol;
Output
    $\int f \, dx$ or the global symbol **Fail**;
Local Variables F, g;
Begin
1    $F := Integral\_table(f, x)$;
2   **if** $F =$ **Fail then**
3        $F := Linear\_properties(f, x)$;
4   **if** $F =$ **Fail then**
5        $F := Substitution\_method(f, x)$;
6   **if** $F =$ **Fail then**
7        $g := Algebraic\_expand(f)$;
8   **if** $f \neq g$ **then**
9            $F := Integral(g, x)$;
10   $Return(F)$
End

Procedure *Substitution_method*(f, x);
Input
    f : an algebraic expression;
    x : a symbol;
Output
    $\int f \, dx$ or the global symbol **Fail**;
Local Variables P, F, i, u, g;
Global v;
Begin
1    $P := Trial\_substitutions(f)$;
2    $F := $ **Fail**;
3    $i := 1$;
4   **while** $F =$ **Fail and** $i \leq Number\_of\_operands(P)$ **do**
5        $g := Operand(P, i)$;
6   **if** $g \neq x$ **and not** $Free\_of(g, x)$ **then**
7            $u := Substitute(f / Derivative(g, x), \; g = v)$;
8   **if** $Free\_of(u, x)$ **then**
9                $F := Substitute(Integral(u, v), \; v = g)$;
10      $i := i + 1$;
11   $Return(F)$
End

Figure 5.17. The MPL *Integral* and *Substitution_method* procedures. (Implementations: Maple (txt), Mathematica (txt), MuPAD (txt).)

	Operator	Integrand	Integration Variable
1	*Integral*	$2\,x\,\cos\left(x^2\right)$	x
2	*Linear_properties*	$2\,x\,\cos\left(x^2\right)$	x
3	*Integral*	$x\,\cos\left(x^2\right)$	x
4	*Substitution_method*	$x\,\cos\left(x^2\right)$	x
5	*Integral*	$(1/2)\,\cos(v)$	v
6	*Linear_properties*	$(1/2)\,\cos(v)$	v
7	*Integral*	$\cos(v)$	v
8	*Integral_table*	$\cos(v)$	v

Figure 5.18. The sequence of procedure calls that contribute to the evaluation of $\int 2\,x\,\cos\left(x^2\right)\,dx$.

calls that return **Fail** and don't contribute to the evaluation.) At step 1, *Integral* calls on *Linear_properties* (step 2) where the leading constant 2 is removed. At step 3, *Linear_properties* passes the new expression $x\,\cos\left(x^2\right)$ to *Integral* which, at step 4, calls on *Substitution_method*. This step introduces a new leading constant $1/2$ and passes a new integrand to *Integral* (step 5). At step 6, the leading constant $1/2$ is removed by another call to *Linear_properties* which again passes a new integrand to *Integral* (step 7). Finally, at step 8, *Integral* calls on *Integral_table* which terminates the recursion and returns $\sin(v)$. At this point the recursion unwinds to give

$$\int 2\,x\,\cos\left(x^2\right)\,dx = \sin\left(x^2\right).$$

□

Appraisal of the Algorithm

The algorithm can evaluate many integrals that depend on the application of the linear properties and the inversion of the chain rule, but cannot evaluate all such integrals. For example, for the integral

$$\int \frac{2x}{x^4+1}\,dx = \arctan\left(x^2\right),$$

the set of possible substitutions obtained by the algorithm is

$$P = \left\{x^4+1, -1, x, 4\right\}.$$

Since this integral is evaluated using the substitution $v = x^2$, which is not in P, the integration is not obtained with the algorithm[5].

---

[5] The reader wishing to explore substitutions of this type should consult Cohen [24], Section 4.4, Exercise 10(d).

In other cases, although the substitution is in P, the algorithm cannot evaluate the integral because of the form of the integrand. For example, consider the integral

$$\int \frac{dx}{\exp(x) + \exp(-x)}.$$

In this form the substitution set is

$$P = \{\exp(x) + \exp(-x), \quad \exp(x), \quad \exp(-x), \quad x, \quad -x\}.$$

Although this integral can be evaluated with $v(x) = \exp(x)$, this substitution will not work with the integrand in this form. However, by multiplying the numerator and denominator of the integrand by $\exp(x)$, we obtain

$$\int \frac{\exp(x)}{(\exp(x))^2 + 1} \, dx = \arctan(\exp(x)),$$

which is evaluated with the substitution $v(x) = \exp(x)$. Since our algorithm does not perform the transformation

$$\frac{1}{\exp(x) + \exp(-x)} \rightarrow \frac{\exp(x)}{(\exp(x))^2 + 1},$$

it cannot evaluate the integral.

Some extensions of the algorithm are described in Exercises 4, 6, 8, and 9.

Exercises

1. For each of the following integrals, give the sequence of procedure calls that shows the path taken by the algorithm to evaluate the integral. For some integrals, the sequence of procedure calls depends on the order of the expressions in the substitution set P.

   (a) $\int \sec^3(x) \tan(x) \, dx.$

   (b) $\int (\sin(x) + 4)^3 \cos(x) \, dx.$

   (c) $\int x \cdot \left(\left(\frac{x^2}{2\,c} + 1 \right)^3 + 2 \right) dx.$

   (d) $\int (\sin(x) + 1)(\cos(x) + 1) \, dx.$

2. Explain why each of the following integrals can be evaluated with substitution but cannot be evaluated by the algorithm in this section.

(a) $\displaystyle\int \frac{x+2}{x^2+4\,x+2}\;dx,\quad$ let $v = x^2 + 4x + 2.$

(b) $\displaystyle\int \sin\left(a\,x^2 + b\,x^2\right)x\;dx,\quad$ let $v = a\,x^2 + b\,x^2.$

(c) $\displaystyle\int \frac{1}{(2\,x+3)\sqrt{4\,x+5)}}\;dx = \arctan(\sqrt{4x+5}),\quad$ let $v = \sqrt{4x+5}.$

3. (a) Give a procedure for *Integral_table*(f, x). If f is not in the table, return the global symbol **Fail**.

(b) Give a procedure for the *Linear_properties*(f, x) operator.

When f is a product, apply Equation (5.15) by separating the operands that are free of x from f using the *Separate_factors* procedure (see page 148) and integrating the remaining expression with a recursive call to *Integral*. If none of the operands of f is free of x, this property does not contribute to the evaluation of the integral, and so the procedure returns the global symbol **Fail**.

When f is a sum, apply Equation (5.16) by evaluating the integral of each operand using *Integral*. However, if some operand cannot be integrated, then return **Fail** because the algorithm cannot integrate the entire sum.

Finally, if f is not a product or a sum, return **Fail**.

(c) Give a procedure *Trial_substitutions*(f) that finds all functions, arguments of functions, and bases and exponents of powers that occur in f. The result should be returned as a set.

4. This exercise describes a procedure that evaluates integrals of rational expressions of the form

$$\int \frac{r\,x + s}{a\,x^2 + b\,x + c}\;dx, \tag{5.28}$$

where $a \neq 0$, b, c, r, and s are free of x. The algorithm for this integral is divided into two cases. First, when the integrand has the form

$$f = \frac{1}{q}, \quad q = a\,x^2 + b\,x + c,$$

then

$$\int \frac{dx}{q} = \begin{cases} 2\dfrac{\arctan\left(\dfrac{2\,a\,x+b}{\sqrt{4\,a\,c-b^2}}\right)}{\sqrt{4\,a\,c-b^2}}, & \text{if } b^2 - 4\,a\,c < 0, \\[3ex] -2\dfrac{\operatorname{arctanh}\left(\dfrac{2\,a\,x+b}{\sqrt{b^2-4\,a\,c}}\right)}{\sqrt{b^2-4\,a\,c}}, & \text{if } b^2 - 4\,a\,c > 0, \\[3ex] -\dfrac{2}{2\,a\,x+b}, & \text{if } b^2 - 4\,a\,c = 0. \end{cases} \tag{5.29}$$

Next, when the integrand has the form $f = \dfrac{r\,x + s}{q}$, then

$$\int \frac{r\,x + s}{q}\,dx = \alpha \ln(q) + \beta \int \frac{dx}{q},$$

where $\alpha = r/(2\,a)$, $\beta = s - r\,b/(2\,a)$, and the integral on the right is evaluated with Equation (5.29).

(a) Give a procedure *Rational_form*(f, x) that checks that f has the proper form and obtains the integral. If $b^2 - 4\,a\,c$ is not an integer or fraction, which means that it cannot be compared to 0, return the arctan form in (5.29). If f does not have the form in (5.28), return the global symbol **Fail**. *Hint:* The *Linear_form* procedure (see Figure 5.7 on page 181) and the *Quadratic_form* procedure (Exercise 8, page 195) are useful for this exercise.

(b) Modify the *Rational_form* procedure so that it also evaluates integrals of the form

$$\int \frac{r\,x + s}{b\,x + c}\,dx,$$

where $b \neq 0$ and $r \neq 0$. (The cases $b = 0$ or $r = 0$ are handled by other cases in *Integral*.)

(c) Modify the main *Integral* procedure so that it calls on *Rational_form*.

(For further exploration of this operator and its generalization, the reader may consult Cohen [24], Section 4.4, Exercise 10.)

5. Use the *Integral* operator together with the *Rational_form* operator in Exercise 4 to evaluate

$$\int \frac{\cos(x)}{\sin^2(x) + 3\sin(x) + 4}\,dx.$$

6. This exercise describes a procedure that evaluates integrals of the form

$$\int \frac{1}{(a\,x + b)\sqrt{r\,x + s}}\,dx,$$

where $a \neq 0$, b, $r \neq 0$, and s are free of x.

(a) Show that the integral can be transformed by the substitution $v = \sqrt{r\,x + s}$ to

$$\int \frac{1}{(a\,x + b)\sqrt{r\,x + s}}\,dx = 2 \int \frac{1}{a\,v^2 - a\,s + b\,r}\,dv.$$

The new integral is evaluated using the *Rational_form* operator (Exercise 4).

(b) Give a procedure *Radical_form*(f, x) that checks if f has the proper form, and, if so, applies the above transformation and returns the result in terms of x. If f does not have the proper form, return the global symbol **Fail**.

(c) Modify the main *Integral* procedure so that it calls on *Radical_form* procedure.

7. Use the *Integral* operator together with the *Radical_form* operator (Exercise 6) to evaluate

$$\int \frac{3\cos(x)}{(5\sin(x)+1)\sqrt{4\sin(x)+7}}\, dx.$$

8. This exercise describes a procedure that evaluates integrals of the form

$$\int f\, dx = \int \sin^m(x)\, \cos^n(x)\, dx, \tag{5.30}$$

where m and n are non-negative integers. Integrals of this form can be evaluated using the reduction formulas

$$\int \cos^n(x)\, dx = (1/n)\cos^{n-1}(x)\, \sin(x) + \frac{n-1}{n}\int \cos^{n-2}(x)\, dx, \tag{5.31}$$

$$\int \sin^m(x)\, \cos^n(x)\, dx \;=\; -\frac{\sin^{m-1}(x)\, \cos^{n+1}(x)}{m+n} \tag{5.32}$$

$$+\frac{m-1}{m+n}\int \sin^{m-2}(x)\, \cos^n(x)\, dx.$$

Notice that repeated use of Equation (5.32) reduces the integrand in (5.30) to the form $\sin(x)\cos^n(x)$ (when m is odd) or to $\cos^n(x)$ (when m is even). In the first case, the remaining integral is evaluated by a call to *Substitution_method* and, in the second case, with Equation (5.31).

(a) Give a procedure *Trig_form*(f, x) that checks if f has the proper form and if so obtains the integral. If f does not have the proper form, return the global symbol **Fail**.

(b) Modify the main *Integral* procedure so that it calls on *Trig_form*.

Another approach for these integrals is described in Exercise 10, page 306.

9. Let n be a positive integer and let a and b be free of x. The following recurrence relations are derived using integration by parts:

$$\int x^n\, \exp(a\,x+b)\, dx = \tag{5.33}$$

$$x^n/a\, \exp(a\,x+b) - n/a\int x^{n-1}\, \exp(a\,x+b)\, dx,$$

$$\int x^n\, \sin(a\,x+b)\, dx = \tag{5.34}$$

$$-x^n/a\, \cos(a\,x+b) + n/a\int x^{n-1}\, \cos(a\,x+b)\, dx,$$

$$\int x^n\, \cos(a\,x+b)\, dx = \tag{5.35}$$

$$x^n/a\, \sin(a\,x+b) - n/a\int x^{n-1}\, \sin(a\,x+b)\, dx.$$

(a) Give a procedure *By_parts*(f, x) that checks if the integrand is one of these forms and when this is so evaluates the integral using the appropriate recurrence relation. If f does not have one of these forms, return the global symbol **Fail**.

(b) Modify the main *Integral* procedure so that it calls on the *By_parts* procedure.

Further Reading

5.1 A Computational View of Recursion. A more detailed discussion of how recursion is implemented in a computer system is given in Pratt [82].

5.2 Recursive Procedures. Rule-based programming in Mathematica is described in Gaylord et al. [38] and Gray [41].

5.3 Case Study: An Elementary Indefinite Integration Operator. Moses [70] discusses the derivative divides method of integration. Symbolic integration is a very difficult mathematical and computational problem. Geddes, Czapor, and Labahn [39], Chapter 11 is a good introduction to the subject. Bronstein [13] gives a theoretical discussion of the subject.

6

Structure of Polynomials and Rational Expressions

In Chapter 3 we described the tree structure of an expression. An expression also has a *semantic structure* that is related to its mathematical properties. For example, the expression $3\,x^2 + 4\,x + 5/2$ can be viewed both as an expression tree and semantically as a polynomial in x with degree 2 that has rational number coefficients.

In this chapter, we describe the *polynomial structure* and *rational expression structure* of an algebraic expression. For polynomials, we give three definitions of increasing generality: first for single variable polynomials (Section 6.1); next for multivariate polynomials (Section 6.1); and finally for general polynomial expressions (Sections 6.2 and 6.3). The definitions are more involved than those found in mathematics textbooks, since they focus on computational concerns as well as the mathematical concept of a polynomial. Along with these definitions, we give MPL procedures that determine the polynomial structure of an expression. In Section 6.4, we use these structural concepts to describe the goals of two transformations, coefficient collection and algebraic expansion, and give MPL algorithms for these operations. Finally, in Section 6.5 we describe the rational expression structure of an algebraic expression and give an algorithm that transforms an expression to a particular rational form.

Although operators that determine the structure of polynomials and rational expressions are available in most computer algebra languages, their capacity varies from system to system. The concepts in this chapter provide a framework to analyze and compare how these concepts are implemented in various CAS languages.

6.1 Single Variable Polynomials

We begin by considering polynomials in a single variable with rational number coefficients.

Definition 6.1. (**Mathematical Definition**) *A* **polynomial** u *in a single variable x is an expression of the form:*

$$u = u_n x^n + u_{n-1} x^{n-1} + \ldots + u_1 x + u_0, \tag{6.1}$$

where the **coefficients** *u_j are rational numbers, and n is a non-negative integer. If $u_n \neq 0$, then u_n is called the* **leading coefficient** *of u and n is its* **degree**. *The expression $u = 0$ is called the* **zero polynomial**; *it has leading coefficient 0 and, according to mathematical convention has degree $-\infty$. The leading coefficient is represented by $\mathrm{lc}(u, x)$ and the degree by $\deg(u, x)$. When the variable x is evident from context, we use the simpler notations $\mathrm{lc}(u)$ and $\deg(u)$.*

Observe that we have distinguished the zero polynomial from other constant polynomials because it has no non-zero coefficients, and so the general definitions for leading coefficient and degree do not apply[1].

Example 6.2.

$$u = 3x^6 + 2\,x^4 - 5/2, \quad \deg(u) = 6, \quad \mathrm{lc}(u) = 3,$$

$$u = x^2 - x + 2, \quad \deg(u) = 2, \quad \mathrm{lc}(u) = 1, \tag{6.2}$$

$$u = 2x^3, \quad \deg(u) = 3, \quad \mathrm{lc}(u) = 2, \tag{6.3}$$

$$u = 3, \quad \deg(u) = 0, \quad \mathrm{lc}(u) = 3. \tag{6.4}$$

$\square$

Although Definition 6.1 defines the concept of a polynomial in a mathematically precise way, it requires some interpretation and is not adequate for computational purposes. For example, in the previous example, the definition is interpreted in a broad sense to include expressions that have coefficients that are understood to be ± 1 (as in Equation (6.2)) and those that have a single term (as in Equations (6.3) and (6.4)). The following definition, which captures the essence of a single variable polynomial in a

---

[1] For the polynomial $u = 0$, both Maple's **degree** operator and Mathematica's **Exponent** operator return a degree of $-\infty$. On the other hand, MuPAD's degree operator returns a degree of 0.

computational setting, can be easily expressed as an MPL procedure that recognizes when an expression is a polynomial.

Definition 6.3. **(Computational Definition)** *A* **monomial** *in a single variable* x *is an algebraic expression* u *that satisfies one of the following rules.*

MON-1. *u is an integer or fraction.*

MON-2. *$u = x$.*

MON-3. *$u = x^n$, where $n > 1$ is an integer.*

MON-4. *u is a product with two operands that satisfies either MON-1, MON-2, or MON-3.*

A **polynomial** *in a single variable* x *is an expression* u *that satisfies one of the following rules.*

POLY-1. *u is a monomial in x.*

POLY-2. *u is a sum, and each operand of u is a monomial in x.*

Primitive Operations on Polynomials

The *Monomial_sv* and *Polynomial_sv* Operators. The operators that are described in the next definition recognize when an expression is a monomial or a polynomial.

Definition 6.4. *Let u be an algebraic expression. The operator*

$$Monomial\_sv(u, x)$$

returns **true** *when u is a monomial in x and otherwise returns* **false**. *(The suffix "sv" stands for "single variable.") The operator*

$$Polynomial\_sv(u, x)$$

returns **true** *when u is a polynomial in x and otherwise returns* **false**.

Example 6.5.

$$
\begin{aligned}
Monomial\_sv(2\,x^3,\ x) &\rightarrow \textbf{true}, \\
Monomial\_sv(x + 1,\ x) &\rightarrow \textbf{false}, \\
Polynomial\_sv(3\,x^2 + 4\,x + 5,\ x) &\rightarrow \textbf{true}, \\
Polynomial\_sv(1/(x + 1),\ x) &\rightarrow \textbf{false}, \\
Polynomial\_sv(a\,x^2 + b\,x + c,\ x) &\rightarrow \textbf{false}.
\end{aligned}
$$

The expression $a\,x^2+b\,x+c$ is not a polynomial in x because the coefficients are not rational numbers. It is, however, a multivariate polynomial (Definition 6.12, page 221), and a general polynomial expression (Definition 6.14, page 223).   □

The operators described in Definition 6.4 are understood to operate within a computational environment defined by an evaluation process that includes automatic simplification. Since this process is applied to the input arguments before the actual tests are done, an expression u is a polynomial in x if the evaluation process transforms it to an expression that satisfies Definition 6.3. In this sense, $\sin(x)+x-\sin(x)$ is a polynomial in x, because the $\sin(x)$ terms are eliminated by automatic simplification.

But now the question arises, should the operators apply any other transformation rules to u before the tests are done? In other words, in what simplification context should we interpret our polynomial definition? For example, each of the expressions

$$(x+1)(x+3), \quad x^2+\sin^2(x)+\cos^2(x), \quad \frac{x^2-1}{x-1}, \quad \cos(2\arccos(x))$$

can be transformed to a polynomial in the sense of Definition 6.1. However, if we assume a simplification context of automatic simplification, they are not considered polynomials in x because the required transformation rules are not applied by this process.

The question of which simplification transformations to include in the definition of *Polynomial_sv* does not have a simple answer. For example, if the operator *Algebraic_expand* were applied, the expression $(x+1)(x+3)$ would be a polynomial in x. There are, however, some cases when it is not useful to apply *Algebraic_expand* (for example, see Expression (6.9) on page 224). For now, we take the conservative view that the these procedures as well as the others in this section operate within the context of only automatic simplification.

Procedures for *Monomial_sv* and *Polynomial_sv* are given in Figures 6.1 and 6.2. In *Monomial_sv*, the four MON tests are done in lines 1-11. Notice that MON-4 (lines 10-11) is handled with two recursive calls on the procedure. Any expression that is not handled by lines 1-11 is not a monomial in x, and so **false** is returned (line 12). In a similar way, *Polynomial_sv* tests the two POLY rules in lines 1-7. Any expression not handled here is not a polynomial, and so **false** is returned at line 8.

The operators in the next three definitions provide a way to analyze the polynomial structure of an expression.

Procedure *Monomial_sv*(u, x);
Input
    u : an algebraic expression;
    x : a symbol;
Output
    true or **false**;
Local Variables
    base, *exponent*;
Begin
1   **if** *Kind*(u) ∈ {**integer**, **fraction**} **then**
2   *Return*(**true**)
3   **elseif** $u = x$ **then**
4   *Return*(**true**)
5   **elseif** *Kind*(u) = " ∧ " **then**
6   *base* := *Operand*(u, 1);
7   *exponent* := *Operand*(u, 2);
8   **if** *base* = x **and** *Kind*(*exponent*) = **integer and** *exponent* > 1 **then**
9   *Return*(**true**)
10   **elseif** *Kind*(u) = " ∗ " **then**
11   *Return*(*Number_of_operands*(u) = 2 **and** *Monomial_sv*(*Operand*(u, 1), x)
                and *Monomial_sv*(*Operand*(u, 2), x));
12   *Return*(**false**)
End

Figure 6.1. An MPL monomial recognition procedure. (Implementation: Maple (txt), Mathematica (txt), MuPAD (txt).)

The *Degree_sv* Operator

Definition 6.6. *Let u be an algebraic expression. If u is a polynomial in x, the operator*

$$Degree\_sv(u, x)$$

returns $\deg(u, x)$. *If u is not a polynomial in x, the operator returns the symbol* **Undefined**.

Example 6.7.

$$Degree\_sv(3x^2 + 4x + 5, \ x) \ \rightarrow \ 2,$$
$$Degree\_sv(2x^3, \ x) \ \rightarrow \ 3,$$
$$Degree\_sv((x + 1)(x + 3), \ x) \ \rightarrow \ \textbf{Undefined},$$
$$Degree\_sv(3, \ x) \ \rightarrow \ 0. \qquad \qquad \square$$

Procedure *Polynomial_sv*(u, x);
Input
   u : an algebraic expression;
   x : a symbol;
Output
   true or **false**;
Local Variables
   i;
Begin
1   **if** *Monomial_sv*(u, x) **then**
2   *Return*(**true**)
3   **elseif** *Kind*$(u) = $ " $+$ " **then**
4   **for** $i := 1$ **to** *Number_of_operands*(u) **do**
5   **if** *Monomial_sv*$(Operand(u, i), x) = $ **false then**
6   *Return*(**false**);
7   *Return*(**true**);
8   *Return*(**false**)
End

Figure 6.2. An MPL polynomial recognition procedure. (Implementation: Maple (txt), Mathematica (txt), MuPAD (txt).)

Procedures for the operator *Degree_sv*(u, x), similar to the ones for *Monomial_sv* and *Polynomial_sv*, are given in Figures 6.3 and 6.4. In this case, the procedure *Degree_monomial_sv*(u, x) gives the degree for monomials, and *Degree_sv*(u, x) is defined in terms of this procedure. Observe that in *Degree_monomial_sv* at line 17, we use the structural assumption that a constant in a product is the first operand (Rule 2, page 90)[2].

The *Coefficient_sv* Operator

Definition 6.8. *Let u be an algebraic expression. If u is a polynomial in x, the operator*

$$Coefficient\_sv(u, x, j)$$

returns the coefficient u_j of x^j in Equation (6.1). Coefficient_sv returns 0, if $j > \deg(u, x)$. If u is not a polynomial in x, the operator returns the symbol **Undefined**.

---

[2]This assumption holds in both Maple and Mathematica. In MuPAD, however, since the constant is the last operand in a product, line 17 is replaced by *Return*(s).

Procedure $Degree\_monomial\_sv(u, x)$;
Input
    u : an algebraic expression;
    x : a symbol;
Output
    $\deg(u, x)$ or the global symbol **Undefined**;
Local Variables
    $base, exponent, s, t$;
Begin

```
1    if  u = 0 then
2        Return(−∞)
3    elseif  Kind(u) ∈ {integer, fraction} then
4        Return(0)
5    elseif  u = x then
6        Return(1)
7    elseif  Kind(u) = ” ∧ ” then
8        base := Operand(u, 1);
9        exponent := Operand(u, 2);
10       if  base = x and Kind(exponent) = integer and exponent > 1 then
11           Return(exponent)
12   elseif  Kind(u) = ” ∗ ” then
13       if  Number_of_operands(u) = 2 then
14           s := Degree_monomial_sv(Operand(u, 1), x);
15           t := Degree_monomial_sv(Operand(u, 2), x);
16           if  s ≠ Undefined and t ≠ Undefined then
17               Return(t)
18   Return(Undefined)
End
```

Figure 6.3. An MPL procedure for *Degree_monomial_sv*. (Implementation: Maple (txt), Mathematica (txt), MuPAD (txt).)

Example 6.9.

$$Coefficient\_sv(x^2 + 3x + 5,\ x,\ 1) \quad \rightarrow \quad 3,$$
$$Coefficient\_sv(2x^3 + 3x,\ x,\ 4) \quad \rightarrow \quad 0,$$
$$Coefficient\_sv(3,\ x,\ 0) \quad \rightarrow \quad 3,$$
$$Coefficient\_sv((x + 1)(x + 3),\ x,\ 2) \quad \rightarrow \quad \textbf{Undefined}. \qquad \square$$

The *Coefficient_sv* operator is implemented with procedures similar to those for *Polynomial_sv* and *Degree_sv* (Exercise 6).

Procedure *Degree_sv(u, x)*;
Input
    u : an algebraic expression;
    x : a symbol;
Output
    deg(*u, x*) or the global symbol **Undefined**;
Local Variables
    d, i, f;
Begin
```
1     d := Degree_monomial_sv(u, x);
2     if  d ≠ Undefined then
3         Return(d)
4     elseif  Kind(u) = " + " then
5         d := 0;
6         for  i := 1 to  Number_of_operands(u) do
7             f := Degree_monomial_sv( Operand(u, i), x);
8             if  f = Undefined then
9                 Return(Undefined)
10            else
11                d := Max({d, f})
12        Return(d);
13    Return(Undefined)
```
End

Figure 6.4. An MPL procedure for *Degree_sv*. (Implementation: Maple (txt), Mathematica (txt), MuPAD (txt).)

The *Leading_coefficient_sv* Operator

Definition 6.10. *Let u be an algebraic expression. If u is a polynomial in x, the operator*

$$Leading\_coefficient\_sv(u, x)$$

returns lc(*u, x*) *(Definition 6.1, page 214). If u is not a polynomial in x, the operator returns the symbol* **Undefined**.

Example 6.11.

$$Leading\_coefficient\_sv(x^2 + 3x + 5, \, x) \quad \rightarrow \quad 1,$$
$$Leading\_coefficient\_sv(3, \, x) \quad \rightarrow \quad 3.$$

□

The *Leading_coefficient_sv* operator can be obtained with a composition of the *Degree_sv* and *Coefficient_sv* operators. For example, if $u = 3x^2 + 4x + 5$, the leading coefficient is obtained with

$$Coefficient\_sv(u, x, Degree\_sv(u, x)) \to 3.$$

Another approach is to obtain it directly with procedures similar to those for *Degree_sv* (Exercise 7).

Multivariate Polynomials

Polynomials that contain more than one variable are called *multivariate polynomials*.

Definition 6.12. (Mathematical Definition) *A* **multivariate polynomial** *u in the set of symbols* $\{x_1, x_2, \ldots, x_m\}$ *is a finite sum with (one or more) monomial terms of the form*

$$c\,x_1^{n_1}\,x_2^{n_2} \cdots x_m^{n_m},$$

where the coefficient c is a rational number and the exponents n_j are non-negative integers.

Example 6.13. The following are multivariate polynomials:

$$p + 1/2\,\rho v^2 + \rho g y, \quad a x^2 + 2 b x + 3 c, \quad x^2 - y^2, \quad m c^2, \quad 3 x^2 + 4. \ \square$$

Although it is possible to give a computational definition for multivariate polynomials that is similar to Definition 6.3 and to extend the primitive operations to this setting (Exercise 2), it is more convenient to do so in the context of general polynomial expressions, which are defined in the next section.

Exercises

For the exercises in this section, do not use the polynomial operators in a CAS.

1. The *height* of a polynomial is the maximum of the absolute values of its coefficients. Let u be an algebraic expression. Give a procedure

$$Polynomial\_height(u, x)$$

   that returns the height of a polynomial. If u is not a polynomial in x, return the global symbol **Undefined**.

2. (a) Give a computational definition for multivariate polynomials that is similar to Definition 6.3.

(b) Give a procedure

$$Polynomial\_mv(u, S)$$

that returns **true** if an algebraic expression u is a multivariate polynomial in a set S of symbols and otherwise returns **false**.

3. Give a definition for a polynomial that includes expressions that contain products and positive integer powers of expressions that satisfy Definition 6.3. For example,

$$x^3 + (x+1)(x+2) + 4, \quad (x^2+x+1)^3, \quad 1+(x+1)(x+2)^2, \quad ((x+1)^2+1)^2$$

are polynomials according to this new definition. Give a procedure

$$Polynomial\_sv\_unexp(u, x)$$

that returns **true** if an algebraic expression is a polynomial in this sense and otherwise returns **false**. Do not use the *Algebraic_expand* operator as part of the definition or the procedure.

4. Consider the class of expressions that are polynomials in y with coefficients that are polynomials in x with rational number coefficients. For example, $u = (1 + x^2)\,y^3 + (2x - 1)\,y$ is in this class. Give a procedure *Polynomial_xy*(u, x, y) that returns **true** if an algebraic expression is in this class and otherwise returns **false**.

5. Consider the class of expressions that are polynomials in x with coefficients that have the form $c + d\sqrt{2}$ where c and d are rational numbers. For example, the expressions $x^3 + (1 - \sqrt{2})\,x^2 + 3 + \sqrt{2}$ and $2\sqrt{2}\,x - 1$ are in this class. Give a procedure *Polynomial_sq2*(u, x) that returns **true** if an algebraic expression is in this class, and otherwise returns **false**.

6. Give a procedure for *Coefficient_sv*(u, x, j). *Hint:* First give a procedure

$$Coefficient\_monomial\_sv(u, x)$$

that returns a list $[c, m]$, where m is the degree of the monomial and c is the coefficient of x^m. If u is not a monomial, return the global symbol **Undefined**.

7. Give a procedure for *Leading_coefficient_sv*(u, x) that does not use the *Degree_sv* or *Coefficient_sv* operators. *Hint:* Modify the procedures for *Degree_sv*. First give a procedure

$$Leading\_coefficient\_monomial\_sv(u, x)$$

that returns a list $[c, m]$, where m is the degree of the monomial and c is the coefficient of x^m. If u is not a monomial, return the global symbol **Undefined**.

8. Let u be an algebraic expression. When u is a polynomial in x, the procedure $Coefficient\_list(u, x)$ returns the list of coefficients of powers of x in u. When u is not a polynomial in x, the procedure returns the global symbol **Undefined**. For example,

$$Coefficient\_list(2\,x^5 + 3\,x^2 + 4\,x + 5, x) \to [2, 3, 4, 5].$$

Give a procedure for $Coefficient\_list(u, x)$.

9. A rational expression in x is an expression of the form p/q, where p and q are polynomials with rational number coefficients. The following are rational expressions:

$$\frac{1}{x + 3}, \qquad \frac{x + 5}{x^2 - 2}, \qquad x^2 - 1,$$

where, in the third example, $q = 1$. Let u be an algebraic expression. Give a procedure

$$Rational\_sv(u, x)$$

that returns **true** when u is a rational expression in x and otherwise returns **false**. Use the numerator and denominator operators in a CAS to obtain p and q (Figure 4.1 on page 124).

6.2 General Polynomial Expressions

There are many expressions that are polynomials in a computational context that are not included in the previous definitions for polynomials. For example, it is reasonable to consider the expression

$$u = \frac{a}{(a + 1)}x^2 + b\,x + \frac{1}{a}$$

as a polynomial in x, even though it does not satisfy the definitions in Section 6.1. Indeed, a CAS views this expression as a polynomial when it solves the quadratic equation $u = 0$ for x. In addition, it is reasonable to view the expressions $\sin^3(x) + 2\,\sin^2(x) + 3$ and $(x + 1)^3 + 2\,(x + 1)^2 + 3$ as polynomials in terms of a complete sub-expression $(\sin(x)$ or $(x + 1))$. On the other hand, the expression $(3\sin(x))x^2 + (2\ln(x))x + 4$ is not a polynomial in x because the coefficients of the powers of x also depend on x.

The next definition includes the more general polynomial expressions given above.

Definition 6.14. (**Mathematical Definition**) Let $c_1, c_2, \ldots, c_r$ be algebraic expressions and let $x_1, x_2, \ldots, x_m$ be algebraic expressions that are *not integers or fractions. A* **general monomial expression** *(GME) in* $\{x_1, x_2, \ldots, x_m\}$ *is an expression of the form*

$$c_1\, c_2 \cdots c_r\, x_1^{n_1}\, x_2^{n_2} \cdots x_m^{n_m}, \tag{6.5}$$

where the exponents n_j *are non-negative integers and each* c_i *satisfies the independence property*

$$\textit{Free\_of}(c_i, x_j) \to \textbf{true}, \quad \textit{for } j = 1, 2, \ldots, m. \tag{6.6}$$

The expressions x_j *are called* **generalized variables** *because they mimic the role of variables, and the expressions* c_i *are called* **generalized coefficients** *because they mimic the role of coefficients. The expression*

$$x_1^{n_1} \cdots x_m^{n_m}$$

is called the **variable part** *of the monomial, and if there are no generalized variables in the monomial, the variable part is 1. The expression* $c_1 \cdots c_r$ *is called the* **coefficient part** *of the monomial, and if there are no generalized coefficients in the monomial, the coefficient part is 1. An expression u is a* **general polynomial expression** *(GPE) if it is either a GME or a sum of GMEs in* $\{x_1, x_2, \ldots, x_m\}$.

Example 6.15. The following are general polynomial expressions:

$$x^2 - x + 1, \quad (x_1 = x),$$
$$x^2\, y - x\, y^2 + 2, \quad (x_1 = x,\ x_2 = y),$$
$$\frac{a}{(a+1)}x^2 + b\, x + \frac{1}{a}, \quad (x_1 = x), \tag{6.7}$$
$$\sin^3(x) + 2\,\sin^2(x) + 3, \quad (x_1 = \sin(x)), \tag{6.8}$$
$$(x+1)^3 + 2\,(x+1)^2 + 3, \quad (x_1 = x+1), \tag{6.9}$$
$$\sqrt{2}\,x^2 + \sqrt{3}\,x + \sqrt{5}, \quad (x_1 = x). \tag{6.10}$$

The definition is quite general. It includes the single variable polynomials (Definition 6.3), multivariate polynomials (Definition 6.12) and allows the more general Expressions (6.7), (6.8), (6.9), and (6.10). Notice that Expression (6.10) is a GPE, but not a single variable polynomial in the sense of Definition 6.1 because the coefficients are not rational numbers. On the other hand, the expression $(\sin(x))\, x^2 + (\ln(x))\, x + 4$ is not a GPE in x alone because the coefficients $\sin(x)$ and $\ln(x)$ do not satisfy the independence property in Equation (6.6).

The definition is also quite flexible because it allows for a choice of which parts of an expression act as variables and which parts act as coefficients.

For example, the expression $2\,a\,x^2 + 3\,b\,x + 4\,c$ can be viewed as a polynomial in $\{a, b, c, x\}$ with integer coefficients or as a polynomial in x with coefficients $2\,a$, $3\,b$ and $4\,c$. In fact, it is possible to view the expression as a polynomial in another variable (say z) with the entire expression as the coefficient part of z^0. In addition, since a sum can be a generalized variable, we can even designate the entire expression as a generalized variable and view it as a polynomial in terms of itself. $\qquad\square$

The following definitions for a GME and GPE are more suitable for computational purposes.

Definition 6.16. (**Computational Definition**) *A* **general monomial expression** *(GME) in a set of generalized variables*

$$S = \{x_1, x_2, \dots, x_m\}$$

is an algebraic expression u that satisfies one of the following rules.

GME-1. *Free_of$(u, x_j) \to$* **true***, for $j = 1, \dots, m$.*

GME-2. *$u \in S$.*

GME-3. *$u = x^n$, where $x \in S$ and $n > 1$ is an integer.*

GME-4. *u is a product, and each operand of u is a GME in S.*

A **general polynomial expression** *(GPE) in a set S of expressions is an algebraic expression u that satisfies one of the following rules.*

GPE-1. *u is a GME in S.*

GPE-2. *u is a sum and each operand of u is a GME in S.*

This definition is similar to Definition 6.3 for single variable polynomials. In this case, however, rule GME-1, which expresses the independence property in Equation (6.6), replaces rule MON-1, which only allows for integers or fractions as coefficients. Although the definition is in terms of a set S of generalized variables, a list L of distinct generalized variables would serve as well. There are a few instances in later sections where we refer to a polynomial in a list of variables.

Primitive Operations for General Polynomial Expressions

The operators described in the following definitions obtain the polynomial structure of an expression.

The *Monomial_gpe* and *Polynomial_gpe* Operators

Definition 6.17. *Let* u *be an algebraic expression, and let* v *be either a generalized variable* x *or a set* S *of generalized variables. The operator*

$$Monomial\_gpe(u, v)$$

returns **true** *whenever* u *is a GME in* $\{x\}$ *or in* S*, and otherwise returns* **false***. The operator*

$$Polynomial\_gpe(u, v)$$

returns **true** *whenever* u *is a GPE in* $\{x\}$ *or in* S*, and otherwise returns* **false***.*

Example 6.18.

$$
\begin{aligned}
Monomial\_gpe(a\,x^2\,y^2, \{x, y\}) &\rightarrow \textbf{true}, \\
Monomial\_gpe(x^2 + y^2, \{x, y\}) &\rightarrow \textbf{false}, \\
Polynomial\_gpe(x^2 + y^2, \{x, y\}) &\rightarrow \textbf{true}, \\
Polynomial\_gpe(\sin^2(x) + 2\,\sin(x) + 3, \sin(x)) &\rightarrow \textbf{true}, \\
Polynomial\_gpe(x/y + 2\,y, \{x, y\}) &\rightarrow \textbf{false}, \\
Polynomial\_gpe((x + 1)\,(x + 3), x) &\rightarrow \textbf{false}. \quad \square
\end{aligned}
$$

Procedures for the operators *Monomial_gpe* and *Polynomial_gpe* are given in Figures 6.5 and 6.6. Although the procedures are based on the rules in Definition 6.16, there are two modifications that are designed to avoid redundant recursive calls on the *Set_free_of* operator. (The *Set_free_of* operator determines if u is free of all of the expressions in a set S (Exercise 1, page 194).) First, in the *Monomial_gpe* procedure the independence property GME-1 is checked at the end of the procedure in line 14 instead of at the beginning. The reason for this has to do with the recursive call in *Monomial_gpe* when u is a product (line 11), together with the recursive nature of *Set_free_of*. If GME-1 were at the beginning of the procedure, the *Set_free_of* operator would test the operands of a product in GME-1, and might need to re-check them again because of the recursive calls on *Monomial_gpe* at line 11. By placing the rule at the end of the procedure we avoid this redundancy.

Next, in the *Polynomial_gpe* procedure, we check rule GPE-1 directly only when u is not a sum (lines 2-3). Since a sum can be a monomial (for example, $u = a + b$ and $S = \{a + b\}$), we check for this possibility separately at line 5. By doing this we avoid redundant calls on *Set_free_of* which would

```
      Procedure  Monomial_gpe(u, v);
      Input
         u : an algebraic expression;
         v : a generalized variable or a set of generalized variables;
      Output
         true or false;
      Local Variables
         i, S, base, exponent;
      Begin
1        if  Kind(v) ≠ set then  S := {v} else  S := v;
2        if  u ∈ S then
3           Return(true)
4        elseif  Kind(u) = "∧" then
5           base := Operand(u, 1);
6           exponent := Operand(u, 2);
7           if  base ∈ S and Kind(exponent) = integer and exponent > 1 then
8              Return(true)
9        elseif  Kind(u) = "*" then
10          for  i := 1 to  Number_of_operands(u) do
11             if  Monomial_gpe(Operand(u, i), S) = false then
12                Return(false);
13          Return(true);
14       Return(Set_free_of(u, S))
      End
```

Figure 6.5. An MPL procedure for the recognition of GMEs. (Implementation: Maple (txt), Mathematica (txt), MuPAD (txt).)

occur if *Monomial_gpe* were used to check if a sum is a monomial and then applied again through *Monomial_gpe* at line 7.

The *Variables* Operator. The polynomial structure of a GPE depends on which expressions are chosen for the generalized variables. The operator in the next definition defines a natural set of generalized variables for an expression.

Definition 6.19. *Let u be an algebraic expression. The operator*

$$Variables(u)$$

is defined by the following transformation rules.

Procedure *Polynomial_gpe*(u, v);
Input
    u : an algebraic expression;
    v : a generalized variable or a set of generalized variables;
Output
    true or **false**;
Local Variables
    i, S;
Begin
1   if *Kind*$(v) \neq$ **set then** $S := \{v\}$ **else** $S := v$;
2   if *Kind*$(u) \neq$ " $+$ " **then**
3   *Return*(*Monomial_gpe*(u, S))
4   **else**
5   if $u \in S$ **then** *Return*(**true**);
6   **for** $i := 1$ **to** *Number_of_operands*(u) **do**
7   if *Monomial_gpe*$(Operand(u, i), S) =$ **false then**
8   *Return*(**false**);
9   *Return*(**true**)
End

Figure 6.6. An MPL procedure for the recognition of GPEs. (Implementation: Maple (txt), Mathematica (txt), MuPAD (txt).)

VAR-1. *If u is an integer or a fraction, then*

$$Variables(u) \to \emptyset.$$

VAR-2. *Suppose u is a power. If the exponent of u is an integer that is greater than 1, then*

$$Variables(u) \to \{Operand(u, 1)\}$$

(the base of u), *otherwise*

$$Variables(u) \to \{u\}.$$

VAR-3. *Suppose u is a sum. Then $Variables(u)$ is the union of the generalized variables of each operand of u obtained using rules VAR-1, VAR-2, VAR-4, or VAR-5.*

VAR-4. *Suppose u is a product. Then $Variables(u)$ contains the union of the generalized variables of each operand of u determined by rules VAR-1, VAR-2, or VAR-5, as well as any operand that is a sum.*

Observe that for a product we include an operand that is a sum in the variable set (see Expression (6.11) below) even though a sum by itself is not in the variable set (VAR-3).

VAR-5. *If u is not covered by the above rules, then*

$$Variables(u) \rightarrow \{u\}.$$

The last rule covers symbols, function forms, and factorials.

Example 6.20. For a multivariate polynomial, the operator returns the set of variables in the expression:

$$Variables(x^3 + 3\,x^2\,y + 3\,x\,y^2 + y^3) \rightarrow \{x, y\}.$$

Other examples include

$$Variables(3\,x\,(x+1)\,y^2 z^n) \rightarrow \{x,\, x+1,\, y,\, z^n\}, \tag{6.11}$$

$$Variables(a\sin^2(x) + 2\,b\sin(x) + 3\,c) \rightarrow \{a,\, b,\, c,\, \sin(x)\},$$

$$Variables(1/2) \rightarrow \emptyset,$$

$$Variables(\sqrt{2}\,x^2 + \sqrt{3}\,x + \sqrt{5}) \rightarrow \{x, \sqrt{2}, \sqrt{3}, \sqrt{5}\}.$$

The last example shows that the *Variables* operator also selects expressions that do not vary in the mathematical sense, but still act as natural place holders in the expression. In fact, any algebraic expression u is always a GPE in terms of *Variables*(u), and when it is viewed in this way, the coefficient part in each monomial is an integer or fraction (Exercise 7). $\square$

The procedure for *Variables*(u) is left to the reader (Exercise 6).

The *Degree_gpe* Operator. In the next definition we generalize the degree concept to generalized polynomial expressions.

Definition 6.21. *Let $S = \{x_1, \ldots, x_m\}$ be a set of generalized variables. Let*

$$u = c_1 \cdots c_r \cdot x_1^{n_1} \cdots x_m^{n_m}$$

*be a monomial with non-zero coefficient part. The **degree** of u with respect to the set S is the sum of the exponents of the generalized variables:*

$$\deg(u, S) = n_1 + n_2 + \cdots + n_m.$$

By mathematical convention, the degree of the 0 *monomial is defined to be* $-\infty$.

If u *is a GPE that is a sum of monomials, then* $\deg(u, S)$ *is the maximum of the degrees of the monomials. If* S *contains a single generalized variable* x, *we use the simpler notation* $\deg(u, x)$, *and if the generalized variables are understood from context, we use* $\deg(u)$.

Example 6.22.

$$\deg(3\,w\,x^2\,y^3\,z^4,\ \{x,\,z\}) = 6,$$

$$\deg(a\,x^2 + b\,x + c,\ x) = 2,$$

$$\deg(a\sin^2(x) + b\sin(x) + c,\ \sin(x)) = 2,$$

$$\deg(2\,x^2\,y\,z^3 + w\,x\,z^6,\ \{x,\,z\}) = 7.\qquad\square$$

Definition 6.23. *Let* u *be an algebraic expression, and let* v *be a generalized variable* x *or a set* S *of generalized variables. The degree operator has the form:*

$$Degree\_gpe(u, v).$$

When u *is a GPE in* v, *the operator returns* $\deg(u, v)$. *If* u *is not a GPE in* v, *the operator returns the global symbol* **Undefined**.

Procedures for *Degree_gpe*, which are similar to the ones for the operators *Monomial_gpe* and *Polynomial_gpe* in Figures 6.5 and 6.6, are left to the reader (Exercise 8).

Definition 6.24. *Let* u *be an algebraic expression, and let*

$$S = Variables(u).$$

The operation $\deg(u, S)$ *is called the* **total degree** *of the expression* u.

Example 6.25. If $u = a\,x^2 + b\,x + c$, then $S = \{a, b, c, x\}$, and the total degree is $\deg(u, \{a, b, c, x\}) = 3$. $\qquad\square$

The *Coefficient_gpe* Operator

Definition 6.26. *Let* u *be an algebraic expression. If* u *is a GPE in a generalized variable* x *and* $j \geq 0$ *is an integer, then the operator*

$$Coefficient\_gpe(u, x, j)$$

returns the sum of the coefficient parts of all monomials of u *with variable part* x^j. *If there is no monomial with variable part* x^j, *the operator returns*

0. If u is not a polynomial in x, the operator returns the global symbol **Undefined**.

Example 6.27.

$$Coefficient\_gpe(a\,x^2 + b\,x + c,\ x,\ 2)\ \rightarrow\ a,$$
$$Coefficient\_gpe(3\,x\,y^2 + 5\,x^2y + 7\,x + 9,\ x,\ 1)\ \rightarrow\ 3\,y^2 + 7,$$
$$Coefficient\_gpe(3\,x\,y^2 + 5\,x^2y + 7\,x + 9,\ x,\ 3)\ \rightarrow\ 0,$$
$$Coefficient\_gpe((3\,\sin(x))\,x^2 + (2\,\ln(x))\,x + 4,\ x,\ 2)\ \rightarrow\ \textbf{Undefined}.$$

$\square$

Procedures that obtain coefficients are shown in Figures 6.7 and 6.8. When u is a GME in x, the *Coefficient_monomial_gpe* procedure returns a list $[c, m]$, where m is the degree u in x and c is the coefficient part of the monomial. If u is not a monomial in x, the global symbol **Undefined** is returned. The case where u is a product is handled in lines 8-18. In lines 9 and 10, the assignments for c and m assume initially that the degree is zero and u is the coefficient part. These values are only changed if some operand of u is a monomial in x with positive degree (lines 15-17). Since u is an automatically simplified product[3], this can happen with at most one operand of u. Nevertheless, we must check each operand to determine that u is a monomial in x.

The *Coefficient_gpe* procedure shown in Figure 6.8 is similar to the *Polynomial_gpe* procedure (Figure 6.6).

Although the *Coefficient_gpe* operator is only defined with respect to a single generalized variable x, the coefficient part of a more general monomial can be obtained by composition. For example, if $u = 3\,x\,y^2 + 5\,x^2y + 7\,x + 9$, the coefficient of $x\,y^2$ can be found with

$$Coefficient\_gpe(\,Coefficient\_gpe(u, x, 1), y, 2)\ \rightarrow\ 3.$$

However, if there are dependencies between the generalized variables, then the order of the coefficient operations is significant. For example, if

$$u = 3\,\sin(x)\,x^2 + 2\,\ln(x)\,x + 4,$$

then to obtain the coefficient of $\ln(x)\,x$, we apply

$$Coefficient\_gpe(\,Coefficient\_gpe(u, \ln(x), 1), x, 1)\ \rightarrow\ 2.$$

---

[3]We assume here that the power transformation $x^m\,x^n \rightarrow x^{n+m}$, for m and n integers, is applied during automatic simplification, and so a product can have at most one operand that is a power with base x. This transformation is obtained in Maple, Mathematica, and MuPAD.

Procedure $Coefficient\_monomial\_gpe(u, x)$;
Input
   u : an algebraic expression;
   x : a generalized variable;
Output
   The list $[c, m]$ where m is the degree of the monomial and
   c is the coefficient of x^m or the global symbol **Undefined**;
Local Variables
   $base, exponent, i, c, m, f$;
Begin

```
1     if  u = x then
2         Return([1, 1])
3     elseif  Kind(u) = " ∧ " then
4         base := Operand(u, 1);
5         exponent := Operand(u, 2);
6         if  base = x and Kind(exponent) = integer and exponent > 1 then
7             Return([1, exponent])
8     elseif  Kind(u) = " * " then
9         m := 0;
10        c := u;
11        for  i := 1  to  Number_of_operands(u) do
12            f := Coefficient_monomial_gpe(Operand(u, i), x);
13            if  f = Undefined then
14                Return(Undefined)
15            elseif  Operand(f, 2) ≠ 0 then
16                m := Operand(f, 2);
17                c := u/x^m;
18        Return([c, m]);
19    if  Free_of(u, x) then
20        Return([u, 0])
21    else
22        Return(Undefined)
```

End

Figure 6.7. An MPL procedure for the $Coefficient\_monomial\_gpe$ operator. (Implementation: Maple (txt), Mathematica (txt), MuPAD (txt).)

The reverse operation

$$Coefficient\_gpe(Coefficient\_gpe(u, x, 1), \ln(x), 1)$$

does not work because the inner coefficient operation returns **Undefined**. It is always possible, however, to order the coefficient operations to deter-

Procedure *Coefficient_gpe*(u, x, j);
Input
    u : an algebraic expression;
    x : a generalized variable;
    j : a non-negative integer;
Output
    The coefficient of x^j in the polynomial u or the global symbol **Undefined**;
Local Variables
    i, c, f;
Begin

```
1     if  Kind(u) ≠ " + " then
2         f := Coefficient_monomial_gpe(u, x);
3         if  f = Undefined then   Return(Undefined)
4         else
5             if  j = Operand(f, 2) then   Return(Operand(f, 1))
6             else  Return(0)
7     else
8         if  u = x then
9             if  j = 1 then   Return(1) else   Return(0);
10        c := 0;
11        for  i := 1 to  Number_of_operands(u) do
12            f := Coefficient_monomial_gpe(Operand(u, i), x);
13            if  f = Undefined then   Return(Undefined)
14            elseif  Operand(f, 2) = j then   c = c + Operand(f, 1);
15        Return(c)
End
```

Figure 6.8. An MPL procedure for the *Coefficient_gpe* operator. (Implementation: Maple (txt), Mathematica (txt), MuPAD (txt).)

mine the desired coefficient (Exercise 8, page 247). Of course, if u is a GPE in both expressions, then the order of the coefficient operations does not matter.

The *Leading_coefficient_gpe* Operator

Definition 6.28. *Let u be an algebraic expression. If u is a GPE in x, then the leading coefficient of u with respect to x is defined as the sum of the coefficient parts of all monomials with variable part $x^{\deg(u,x)}$. The leading coefficient is represented by $\mathrm{lc}(u, x)$, and when x is understood from context, by the simpler notation $\mathrm{lc}(u)$.*

For example, $\mathrm{lc}(3\,x\,y^2 + 5\,x^2 y + 7\,x^2\,y^3 + 9, x) = 5\,y + 7\,y^3$.

Definition 6.29. *Let u be a* GPE *in x. The operator*

$$Leading\_coefficient\_gpe(u, x)$$

returns $lc(u, x)$. *If u is not a* GPE *in x, the operator returns* **Undefined**.

   Leading_coefficient_gpe can be obtained by composition of the operators *Degree_gpe* and *Coefficient_gpe* or directly with procedures similar to those for *Degree_gpe* (Exercise 11).

An Appraisal of the GPE

General polynomial expressions have been defined so that many expressions that are clearly polynomials are included in the definition and many expressions that are not are excluded. In addition, the primitive operators associated with the definition work well in most contexts where it is necessary to examine the polynomial structure of an expression. Since some computer algebra systems have similar operators, the definitions are a good starting point to evaluate and compare the polynomial capacity of CAS software (Exercise 1).

   However, the definition has some limitations. The limitations are associated with the restricted notion of coefficient independence as expressed by the *Free_of* operator in Equation (6.6) and are magnified by the simplification context of automatic simplification in which the operators perform. The following examples illustrate these points.

Example 6.30. Consider the following operations:

$$Polynomial\_gpe(x\,(x^2 + 1),\, x) \rightarrow \textbf{false}, \tag{6.12}$$

$$Polynomial\_gpe(y^2\,(y^4 + 1),\, y^2) \rightarrow \textbf{true}. \tag{6.13}$$

In Expression (6.12), the *Polynomial_gpe* operator concludes that the expression is not a polynomial in x. In this instance the coefficient of x (the expression $x^2 + 1$) is not free of x. Expression (6.13) is obtained from Expression (6.12) with the structural substitution $x = y^2$. In this case, however, the expression is a polynomial in y^2. This follows because y^2 is not a complete sub-expression of $y^4 + 1$ which means that $y^4 + 1$ can act as a coefficient of y^2. The problem here has to do with the limited view of the coefficient part of a monomial which is based on the actions of the *Free_of* operator together with the actions of automatic simplification. $\square$

Example 6.31. Consider the expression $u = a\,(x^2 + 1)^2 + (x^2 + 1)$ which we want to consider as a polynomial in $x^2 + 1$. However,

$$Polynomial\_gpe(u,\ x^2 + 1) \to \textbf{true}, \tag{6.14}$$

$$Degree\_gpe(u,\ x^2 + 1) \to 2, \tag{6.15}$$

$$Coefficient\_gpe(u,\ x^2 + 1,\ 1) \to 0, \tag{6.16}$$

$$Coefficient\_gpe(u,\ x^2 + 1,\ 0) \to x^2 + 1. \tag{6.17}$$

In this case the simplified form of u is

$$a\,(x^2 + 1)^2 + x^2 + 1. \tag{6.18}$$

At this point the expression is still a GPE in $x^2 + 1$ with degree 2 (Expressions (6.14) and (6.15)), although some of its polynomial structure has been changed (Expressions (6.16) and (6.17)). The problem here is the sum on the right in Expression (6.18) $x^2 + 1$ is no longer a complete subexpression of the entire polynomial. Since both x^2 and 1 are free of $x^2 + 1$, they are relegated to the role of a coefficient even though the expression u is a polynomial in the complete sub-expression $x^2 + 1$. $\square$

Example 6.32. Consider the expression $u = 2\,(x^2)^2 + 3\,(x^2)$ and consider the operations:

$$Polynomial\_gpe(u,\ x^2) \to \textbf{true},$$

$$Degree\_gpe(u,\ x^2) \to 1,$$

$$Coefficient\_gpe(u,\ x^2,\ 2) \to 0,$$

$$Coefficient\_gpe(u,\ x^2,\ 1) \to 3,$$

$$Coefficient\_gpe(u,\ x^2,\ 0) \to 2\,x^4. \tag{6.19}$$

Automatic simplification transforms u to $2\,x^4 + 3\,x^2$ which is still a polynomial in x^2, but now the degree is 1, and the polynomial structure has been changed. Indeed, since x^4 is free of x^2 it is considered a coefficient part in Expression (6.19). $\square$

What can we conclude from these examples? First, the definition works best when the generalized variables are symbols, function forms or factorials. When the generalized variables are restricted to expressions of this type, general polynomial expressions are similar to multivariate polynomials with coefficients that are more involved expressions.

The definitions are less reliable, however, when a generalized variable is a sum because the polynomial structure of the expression may be altered by automatic simplification (see Example 6.31). One solution, of course, is simply to modify the GPE definition so that a generalized variable cannot be a sum. We have resisted doing this because the current definition is useful for defining the actions of the *Algebraic_expand* operator in Section 6.4.

The situation is even more discouraging when a generalized variable is a power (see Example 6.32 above) or a product (Exercise 2). Although, we rarely get satisfactory results in these cases, we have included them in the definition because there are a few instances when the primitive operators described here give satisfactory results.

Extensions of the Basic Definitions and Procedures

There are a number of ways that we can remove some of the limitations of our polynomial model. First, we could restrict the class of generalized variables to expressions that are in the set *Variables(u)*. It is noteworthy that the set *Variables(u)* never contains a product or a power with integer exponent ≥ 2, although it can contain a sum. Another possibility is to extend the capabilities of the operators by performing a more involved analysis of an expression. (For a more detailed discussion of this extension, consult Cohen [24], Sections 4.1 and 6.2.)

There are two other extensions of the polynomial model that require only minor modifications to the definitions and basic procedures. One possibility is to allow generalized variables to have negative integer exponents even though this is not ordinarily done in mathematical definitions for polynomials. For example, when this is done the expression $2/x + 3/x^2$ is a polynomial with $\deg(u) = -1$. When negative exponents are allowed, it is also useful to define the operation $\text{low\_deg}(u, x)$ that returns the lowest power of x in the expression. For example, $\text{low\_deg}(2/x + 3/x^2, x) = -2$.

A particularly useful modification of the model is to drop the independence *Free_of* condition GME-1 in Definition 6.16. When this is done an expression such as $(x + 1) x^2 + \ln(x) x + \sin(x)$ is a polynomial in x even though the coefficients $x + 1$, $\ln(x)$, and $\sin(x)$ are not free of x. Although this modification causes expressions to lose some of their polynomial structure, it does allow the degree and coefficient operations to be applied in some useful situations. The definitions and procedures for the basic operators for this model are described in Exercise 12.

Exercises

1. In this exercise we ask you to explore the polynomial capabilities of a CAS. For a CAS, consider its versions of MPL's *Polynomial_gpe*, *Degree_gpe* and *Coefficient_gpe* operators.

(a) In Maple consider the **type** command with **polynom** option, and the **degree** and **coeff** operators.

(b) In Mathematica consider the operators **PolynomialQ**, **Exponent**, and **Coefficient**.

(c) In MuPAD consider the **type** operator with the **PolyExpr** option, and the **degree** and **coeff** operators.

Consider the following questions.

(a) Does the polynomial model in a CAS employ the same coefficient independence condition as the MPL model?

(b) Are sums, products, or powers permitted as generalized variables?

(c) Are negative integer exponents permitted in the polynomial model?

(d) Is expansion part of the simplification context?

(e) Does the model extend the MPL model in significant way?

2. Explain why it is usually not meaningful to view an expression u as a polynomial in terms of an expression v that is a product.

3. Let u be and algebraic expression and let S be a set of generalized variables. Give a procedure

$$Coeff\_var\_monomial(u, S)$$

that returns a two element list with the coefficient part and variable part of u. If u is not a GME in S, the procedure returns the global symbol **Undefined**. (This procedure is used in the *Collect_terms* procedure (see Figure 6.9 on page 249).)

4. Give a procedure

$$Bilinear\_form(u, x, y)$$

that returns **true** when an algebraic expression u has the form $a\,x + b\,y + c$, where x and y are symbols and a, b, and c are free of x and y. If u does not have this form, return **false**. Interpret the form broadly to allow $2\,x + c\,x + 3\,y + d\,y + 4$ to be in this form.

5. What is returned by the *Variables* operator, and what is the total degree for each of the following?

(a) $(x+1)(x+2) + (x+3)$.

(b) $\dfrac{(x+1)^2}{(1-x)^2}$.

(c) $\dfrac{x^2}{a^2} + \dfrac{x}{a} + b$.

(d) $x^m + \sin(x)\,x + 1/(x\,y)$.

6. Let u be an algebraic expression. Give a procedure for *Variables*(u).

7. Suppose an algebraic expression u is viewed as a GPE in $Variables(u)$. Explain why the coefficient part of a monomial in u must be an integer or a fraction.

8. Let u be an algebraic expression, and let v be a generalized variable or a set of generalized variables. In this exercise we ask you to give procedures to compute $\deg(u, v)$ (see Definition 6.23). First, give a procedure

$$Degree\_monomial\_gpe(u, v)$$

that finds the degree of a monomial and then a procedure $Degree\_gpe(u, v)$.

9. Let u be an algebraic expression. Give a procedure $Total\_degree(u)$ that returns the total degree of u. To make the exercise interesting, do not use the $Degree\_gpe$ or $Variables$ operators.

10. Let u be a multivariate polynomial in x and y with rational number coefficients. The polynomial is called a *homogeneous* polynomial if every monomial term has the same total degree. For example, the polynomial $u = x^2 + 2xy + y^2$ is homogeneous. Give a procedure

$$Homogeneous\_polynomial(u, x, y)$$

that returns **true** if u is homogeneous in x and y and **false** otherwise.

11. Let u be an algebraic expression, and let x be a generalized variable.

    (a) Give a procedure for $Leading\_coefficient\_gpe(u, x)$ that does not use the $Degree\_gpe$ or $Coefficient\_gpe$ operators. If u is not a GPE in x, return the global symbol **Undefined**.

    (b) Give a procedure $Leading\_coeff\_degree\_gpe(u, x)$ that returns the list

    $$[\mathrm{lc}(u, x), \deg(u, x)].$$

    If u is not a GPE in x, return the global symbol **Undefined**. Do not use the $Degree\_gpe$ or $Coefficient\_gpe$ operators in this exercise.

12. Let x be a symbol. In this exercise we give an alternate definition of a polynomial that does not require the coefficients of powers of x to be free of x.

    An *algebraic expression* u is an **alternate general monomial expression** in x if it satisfies one of the following rules.

    GMEALT-1. $u = x$.

    GMEALT-2. $u = x^n$ where $n > 1$ is an integer.

    GMEALT-3. u is a product, and each operand of u is either a sum or satisfies one of the rules GMEALT-1, GMEALT-2, or GMEALT-4.

    GMEALT-4. u is an *algebraic expression* that is not sum and does not satisfy rules GMEALT-1, GMEALT-2, or GMEALT-3.

Rules GMEALT-1 and GMEALT-2 give monomials of positive degree, and rule GMEALT-3 gives a monomial of positive degree if one of the operands satisfies GMEALT-1 or GMEALT-2 and otherwise has degree 0. Rule GMEALT-4 gives monomials with degree 0 in x such as

$$2, \quad 3/2, \quad a^2, \quad \sin(x), \quad 1/(x+1).$$

In addition, according to this rule a sum is not a monomial of degree 0. On the other hand, in GMEALT-3 a sum is allowed as an operand in a product that is a monomial. This means that the following are monomials in x:

$$(a+1)\,x^2, \quad (x+1)\,x^2, \quad (x+1)\,(x+2), \quad (x+1)^2,$$

where the first two expressions have degree 2 in x and the last two expressions have degree 0.

An *algebraic expression* u is **alternate general polynomial expression** in x if it satisfies one of the rules:

GPEALT-1. u is an alternate general monomial expression in x.

GPEALT-2. u is a sum, and each operand of u is an alternate general monomial expression in x.

With this definition

$$u = \frac{x^2}{x+1} + \sin(x)\,x + c$$

is a polynomial in x and operations such as $\deg(u, x) = 2$ are well defined. Give procedures

$$Degree\_alternate(u, x), \quad Coefficient\_alternate(u, x, j)$$

that obtain the degree and coefficient operations in this context.

13. Let u be an algebraic expression, and let x be a symbol. In this exercise we extend the basic definitions for monomials and polynomials so that exponents of x can be any algebraic expressions that are free of x. For example, in this context the expression

$$u = x^{m+1} + 3\,x^n + 4\,x^3 + 5\,x + 6\,x^{-1}$$

is a polynomial in x. Give a procedure $Degree\_general(u, x)$ that obtains the degree of these polynomial expressions. If u is not a polynomial in this sense (e.g., x^x), then return the global symbol **Undefined**. For the above polynomial u,

$$Degree\_general(u, x) \rightarrow Max(\{m+1, n, 3\}).$$

In other words, in instances where $Degree\_general$ is unable to actually find the maximum, an unevaluated Max function form is returned. A version of *Max* that obtains this operation is described in Exercise 16 on page 198.

The Mathematica **Exponent** operator performs an operation similar to the one described here.

14. The first step in the *Solve_ode* algorithm (see Figure 4.16 on page 162) transforms a differential equation to the form

$$M + N \frac{dy}{dx} = 0. \tag{6.20}$$

This operation is performed by the *Transform_ode* procedure which returns a list $[M, N]$. The *Transform_ode* assumes that the equation can be transformed to this form but does not check that this is so. Modify the *Transform_ode* procedure so that it checks that the transformed equation has the form of Equation (6.20), where M and N do not contain the function form named d, and when this is so returns $[M, N]$. If the expression cannot be transformed to this form, return the symbol **Fail**. For example,

$$Transform\_ode(x^2 = d(y, x) + y, x, y) \quad \rightarrow \quad [x^2 - y, -1],$$
$$Transform\_ode(x^2 = d(y, x, 2), x, y) \quad \rightarrow \quad \textbf{Fail}.$$

The procedures described in this section and the *Derivative_order* procedure (see Exercise 15 on page 197) are useful for this problem.

15. (a) A *linear differential operator* is an expression of the form

$$a_n \frac{d^n y}{dx^n} + a_{n-1} \frac{d^{n-1} y}{dx^{n-1}} + \cdots + a_1 \frac{dy}{dx} + a_0 y + f$$

where a_i and f are algebraic expressions that are free of y. Let u be an algebraic expression. Give a procedure

$$Linear\_derivative\_order(u, x, y)$$

that determines if u is a linear differential operator and, when this is so, returns the order of the highest derivative of y in u. If u is linear in y but contains none of its derivatives, the procedure returns 0. If u does not contain y or its derivatives, the procedure returns -1. If u is not a linear differential operator, the procedure returns the global symbol **Undefined**. As in Section 4.3, represent the derivative

$$\frac{dy}{dx}$$

with the function notation $d(y, x)$ and higher order derivatives

$$\frac{d^n y}{dx^n}$$

with $d(y, x, n)$. To simplify matters, if u contains any function forms with the name d that contain operands different from those in $d(y, x)$ and $d(y, x, n)$, return the symbol **Undefined**. For example,

$$Linear\_derivative\_order(d(y, x, 2) + 2x, \, x, \, y) \rightarrow 2,$$
$$Linear\_derivative\_order(2\,y + 3\,x, \, x, \, y) \rightarrow 0,$$
$$Linear\_derivative\_order(2\,x + 3, \, x, \, y) \rightarrow -1,$$
$$Linear\_derivative\_order(d(y, x) + y^2, \, x, \, y) \rightarrow \textbf{Undefined},$$
$$Linear\_derivative\_order(d(b), \, x, \, y) \rightarrow \textbf{Undefined}.$$

(b) In Exercise 7, page 169 we describe a procedure $Solve\_ode\_2(a, b, c, f)$ that obtains a solution to the differential equation

$$a\frac{d^2y}{dx^2} + b\frac{dy}{dx} + c\,y = f, \tag{6.21}$$

where a, b, and c are rational numbers and f is an algebraic expression that is free of y. Give a procedure $Transform\_ode\_2(w, x, y)$ that determines if an equation w can be transformed to the form of Equation (6.21) by rational simplification and if so returns the list $[a, b, c, f]$. If w cannot be transformed to this form, then return **Fail**. *Hint:* This procedure is similar to the procedure $Transform\_ode$ in Exercise 14 above. For example,

$$Transform\_ode\_2\,(2\,d(y, x, 2) + 3\,y = x^2, \, x, \, y) \rightarrow [2, 0, 3, x^2],$$
$$Transform\_ode\_2\left(\frac{x^2}{d(y, x, 2)} - 3 = 0, \, x, \, y\right) \rightarrow [-3, 0, 0, -x^2],$$
$$Transform\_ode\_2\,(x\,d(y, x, 2) + 3\,y = x^2, \, x, \, y) \rightarrow \textbf{Fail}.$$

16. A differential equation that has the form

$$\frac{dy}{dx} = P\,y + Q\,y^n \tag{6.22}$$

where $P \neq 0$ and $Q \neq 0$ are free of y and $n \neq 1$ is free of x and y is called a *Bernoulli* equation. For example

$$\frac{dy}{dx} = y + x\,y^3$$

is a Bernoulli equation. This equation is solved by defining a new variable

$$z = y^{1-n} \tag{6.23}$$

that transforms the equation to

$$\frac{1}{1-n}\frac{dz}{dx} = P\,z + Q. \tag{6.24}$$

Once we solve this equation, we obtain the solution to Equation (6.22) with the substitution in Expression (6.23).

(a) Show that Equation (6.24) can be solved using the algorithm in Section 4.3.

(b) Give a procedure $Bernoulli(u, x, y)$ that tries to determine if a differential equation u is a Bernoulli equation and if so uses the procedure $Solve\_ode$ (see Figure 4.16 on page 162) to find the solution to Equation (6.24) and then obtains the solution in terms of y with the substitution in Expression (6.23).

6.3   Relationships Between Generalized Variables[4]

In this section we state and prove a number of mathematical properties of the $Free\_of$ operator and use these properties to investigate the independence of generalized variables.

Mathematical Properties of the $Free\_of$ Operator

Theorem 6.33. *Let u, v, and w be mathematical expressions.*

1. *If $u \neq v$, then $(Free\_of(u, v)$ or $Free\_of(v, u)) \rightarrow$* **true**.

2. **(Transitive Property)** *If $Free\_of(u, v) \rightarrow$* **false** *and $Free\_of(v, w) \rightarrow$* **false**, *then $Free\_of(u, w) \rightarrow$* **false**.

Proof: Both statements are easily proved. To show (1), if $Free\_of(u, v)$ and $Free\_of(v, u)$ are both **false**, then v is a complete sub-expression of u, and u is a complete sub-expression of v. The only way this can happen is for $u = v$. However, $u \neq v$, and so either $Free\_of(u, v)$ or $Free\_of(v, u)$ must be **true**. (Of course, both can be **true**.)

To show (2), the hypothesis states that v is a complete sub-expression of u, and w is a complete sub-expression of v. Therefore, w is a complete sub-expression of u and $Free\_of(u, w) \rightarrow$ **false**.   □

The next theorem extends Theorem 6.33(1) to a set S of expressions.

Theorem 6.34. *Let $S = \{x_1, x_2, \ldots, x_m\}$ be a set of mathematical expressions (with $m \geq 2$). Then, there is an x_j in S such that*

$$Free\_of(x_k, x_j) \rightarrow \textbf{true}, \quad \text{for } k = 1, 2, \ldots, j - 1, j + 1, \ldots, m.$$

---

[4] This section is more theoretical than the previous sections.

Proof: The theorem is proved with mathematical induction. First, for the base case $m = 2$, the theorem follows from Theorem 6.33(1). Next, suppose the theorem is **true** for $S \sim \{x_m\}$. This implies that there is an x_j, with $1 \leq j \leq m - 1$, such that

$$Free\_of(x_k, x_j) \to \textbf{true}, \quad \text{for } k = 1, 2, \ldots, j - 1, j + 1, \ldots, m - 1. \quad (6.25)$$

To show that the theorem holds for S, we consider the two cases where $Free\_of(x_m, x_j)$ is either **true** or **false**. In the first case, $Free\_of(x_m, x_j)$ is **true**, and this assumption, together with the induction hypothesis (6.25), implies that x_j satisfies the theorem for S as well.

For the second case, we assume that

$$Free\_of(x_m, x_j) \to \textbf{false}, \quad\quad\quad\quad (6.26)$$

and show that x_m satisfies the conclusion of the theorem. First, Theorem 6.33(1) applied to Expression (6.26) implies that

$$Free\_of(x_j, x_m) \to \textbf{true}. \quad\quad\quad\quad (6.27)$$

To complete the proof we must show that

$$Free\_of(x_k, x_m) \to \textbf{true}, \quad \text{for } k = 1, 2, \ldots, j - 1, j + 1, \ldots, m - 1. \quad (6.28)$$

However, if for some k, $Free\_of(x_k, x_m)$ were **false**, then this fact, together with Expression (6.26), would imply that $Free\_of(x_k, x_j)$ is **false** which contradicts the induction hypothesis (6.25). Therefore, (6.28), together with (6.27), shows that x_m satisfies the conclusion of the theorem.   □

Example 6.35. If $x_1 = \sin(x)$, $x_2 = \ln(\sin(x))$, and $x_3 = x$, then $j = 2$ and

$$
\begin{aligned}
Free\_of(x_1, x_2) &= Free\_of(\sin(x), \ln(\sin(x))) \to \textbf{true}, \\
Free\_of(x_3, x_2) &= Free\_of(x, \ln(\sin(x))) \to \textbf{true}. \quad\quad □
\end{aligned}
$$

Theorem 6.36. *Let $S = \{x_1, x_2, \ldots, x_m\}$ be a set of (distinct) mathematical expressions. Then, there is a permutation (reordering) of S,*

$$[x_{j_1}, x_{j_2}, \ldots, x_{j_m}]$$

such that (for $i < m$)

$$Free\_of(x_k, x_{j_i}) \to \textbf{true}, \quad \text{for } k = j_{i+1}, \ldots, j_m.$$

The involved notation makes the theorem seem more complicated than it is. It simply states that we can rearrange the expressions in S into a list so that any expression in the list is free of all expressions that precede it in the list.

Proof: [Theorem 6.36] Let the first expression in the list x_{j_1} be the expression described in the conclusion of Theorem 6.34. Then all expressions x_k in $S \sim \{x_{j_1}\}$ satisfy the property *Free_of*$(x_k, x_{j_1}) \to$ **true**. In general, define x_{j_i} to be the expression from the set $S \sim \{x_{j_1}, \ldots, x_{j_{i-1}}\}$ that satisfies the conclusion of Theorem 6.34. $\qquad \Box$

Example 6.37. Suppose $x_1 = \sin(x)$, $x_2 = x$, and $x_3 = \ln(\sin(x))$. A reordering that satisfies the conclusion of the theorem is $i_1 = 3, i_2 = 1, i_3 = 2$ or
$$[\ln(\sin(x)), \sin(x), x].$$
In other words
$$
\begin{aligned}
\textit{Free\_of}(x_{i_2}, x_{i_1}) &= \textit{Free\_of}(\sin(x), \ln(\sin(x))) \to \textbf{true}, \\
\textit{Free\_of}(x_{i_3}, x_{i_1}) &= \textit{Free\_of}(x, \ln(\sin(x))) \to \textbf{true}, \\
\textit{Free\_of}(x_{i_3}, x_{i_2}) &= \textit{Free\_of}(x, \sin(x)) \to \textbf{true}.
\end{aligned}
$$
$\qquad \Box$

In Exercise 2 we describe a procedure that finds the permutation of the expressions guaranteed by Theorem 6.36.

Relationships Between Generalized Variables

Although the definition of a GPE requires the coefficients c_i be independent of the generalized variables x_j, it does not require the generalized variables be independent of each other. For example, the expression

$$2\,x\,(\ln(x))^2 + 3\,x^2\,\ln(x) + 4 \qquad (6.29)$$

is a polynomial in $\{x, \ln(x)\}$, even though the two expressions are not independent (e.g., $\ln(x)$ depends on x). When a dependence relationship like this exists, it is not possible to view the expression as a polynomial in one of the generalized variables. For example, although the expression (6.29) is a polynomial in $\{x, \ln(x)\}$ and in $\ln(x)$ alone, it is not a polynomial in x alone. In this regard, we have the following two theorems.

Theorem 6.38. *If u is a GPE in each of the expressions $x_1, x_2, \ldots, x_m$ individually, then it is also a GPE in $\{x_1, x_2, \ldots, x_m\}$.*

Theorem 6.39. *Suppose u is a GPE in $S = \{x_1, x_2, \ldots, x_m\}$, and suppose that for some j*

$$Free\_of(x_k, x_j) \to true, \quad k = 1, 2, \ldots, m, \quad k \neq j.$$

Then u is a GPE in x_j alone.

The proofs of the theorems are left to the reader (Exercises 4 and 5).

However, if u is a polynomial in a set of expressions, then it must also be a polynomial in at least one of the generalized variables.

Theorem 6.40. *Let u be a GPE in the expressions $S = \{x_1, x_2, \ldots, x_m\}$. Then u is also a GPE in some x_j.*

Proof: This theorem follows from Theorem 6.34 and Theorem 6.39. □

Sometimes it is useful to replace the generalized variables in an expression by symbols. This can be done with concurrent substitution. Let u be a GPE in $S = \{x_1, x_2, \ldots, x_m\}$ and let $y_1, y_2, \ldots, y_m$ be unassigned symbols. The substitution

$$Concurrent\_substitute(u, \ [x_1 = y_1, x_2 = y_2, \ldots, x_m = y_m])$$

creates a multivariate polynomial with each generalized variable x_i replaced by a symbol y_i.

With sequential substitution, however, we may not obtain the intended substitution. This point is illustrated in the next example.

Example 6.41. Consider the expression $u = s \sin(s) \ln(\sin(s))$ as a polynomial in $S = \{s, \sin(s), \ln(\sin(s))\}$. We obtain a multivariate polynomial with the substitutions

$$Sequential\_substitute(u, \ [x = \ln(\sin(s)), \ y = \sin(s), \ z = s]) \to z \, y \, x.$$

On the other hand, if the order of substitutions is

$$Sequential\_substitute(u, \ [y = \sin(s), \ z = s, \ x = \ln(\sin(s))]) \to z \, y \, \ln(y),$$

we don't eliminate all generalized variables. □

However, Theorem 6.36 implies that we can find a re-ordering of the generalized variables $[x_{j_1}, x_{j_2}, \ldots, x_{j_m}]$ so that the substitution

$$Sequential\_substitute(u, \ [x_{j_1} = y_1, x_{j_2} = y_2, \ldots, x_{j_m} = y_m])$$

creates a multivariate polynomial with each generalized variable x_{j_i} replaced by a symbol y_i.

By substituting symbols for generalized variables, we alter some of the polynomial structure of an expression. For example, the new expression will be a GPE in each y_j individually even though the original polynomial may not be a GPE in terms of each of the generalized variables.

Exercises

1. Suppose
$$Free\_of(u, v) \rightarrow \textbf{true}, \qquad Free\_of(v, w) \rightarrow \textbf{true}.$$
Does this imply that $Free\_of(u, w) \rightarrow \textbf{true}$?

2. Let $S = \{x_1, x_2, \ldots, x_m\}$ be a set of (distinct) mathematical expressions. Give a procedure $Free\_of\_sort(S)$ that returns the list of re-ordered expressions described in Theorem 6.36. *Hint:* Any elementary sorting algorithm (insertion sort, bubble sort, selection sort) will do where v precedes u in the list if $Free\_of(u, v) \rightarrow \textbf{true}$.

3. Let S be a set of symbols, and suppose u is a GPE in S. Show that u is a polynomial in each member of S.

4. Prove Theorem 6.38.

5. Prove Theorem 6.39.

6. Let u be an algebraic expression, and consider u as a GPE in $Variables(u)$. Give a procedure $GPE\_to\_mult(u)$ that transforms u to a multivariate polynomial. *Note:* This problem requires an arbitrary number of variable names. The CAS must have the capability to generate variable names[5] or provide subscripted variables.

7. (a) Let L be a list of symbols and let u be a multivariate polynomial in the symbols of L with rational number coefficients. The operator
$$Leading\_numer\_coeff(u, L)$$
obtains the leading numerical coefficient of an expression which is defined using the following rules.

LNC-1. If $L = [\,]$, then
$$Leading\_numer\_coeff(u, L) \rightarrow u.$$

LNC-2. Let $x = First(L, 1)$ and $l = Leading\_coefficient\_gpe(u, x)$. Then,
$$Leading\_numer\_coeff(u, L) \rightarrow Leading\_numer\_coeff(l, Rest(L)). \qquad (6.30)$$

---

[5]For example, in Mathematica the `Unique` command creates symbol names, or in Maple the `cat` command concatenates variable names and integer values. (Implementation: Maple (mws), Mathematica (nb).)

For example,

$$Leading\_numer\_coeff\,(2\,x^2y + 3\,x\,y^2, [x, y]) \to 2,$$

$$Leading\_numer\_coeff\,(2\,x^2y + 3\,x\,y^2, [y, x]) \to 3.$$

Notice that the leading numerical coefficient depends on the order of the symbols in L. Give a procedure for $Leading\_numer\_coeff\,(u, L)$.

(b) Let $s = Leading\_numer\_coeff\,(u, L)$. If $u \neq 0$, define the *polynomial sign* of u with respect to L as the sign $(1$ or $-1)$ of s. In addition define the polynomial sign of 0 as 0. Give a procedure for $Polynomial\_sign(u, L)$ that returns the polynomial sign.

(c) Give a procedure $Polynomial\_sign\_var(u)$ which obtains the polynomial sign of u with respect to the expressions in $Variables(u)$. For this operation it is necessary to replace the generalized variables in u with symbols because the coefficient computation in (6.30) may create new generalized variables (Exercise 6). For example, this happens with $u = -c*(x+y)$ which has the generalized variables c and $x+y$. However, by automatic simplification the coefficient of c is $-x - y$ which has two new generalized variables x and y.

Since the polynomial sign depends on the order of the expressions in the set $Variables(u)$, it is useful to create a list of the generalized variables in a standard order. (For more information on an order relation that can be used for this purpose, consult Cohen [24], Section 3.1.)

8. Suppose that u is a GPE in $S = \{x_1, x_2, \ldots, x_m\}$, and let $n_1, \ldots n_m$ be non-negative integers.

(a) Explain why it is always possible to obtain the coefficient of

$$x_1^{n_1} x_2^{n_2} \cdots x_m^{n_m}$$

using a composition of $Coefficient\_gpe$ operators.

(b) Give a procedure $Coefficient\_vars(u, L)$ where

$$L = [[x_1,\ n_1],\ [x_2, n_2], \ldots, [x_m, n_m]]$$

that implements the statement in part (a).

6.4   Manipulation of General Polynomial Expressions

In this section we describe two operators that manipulate general polynomial expressions. Both of the operators are based on the two distributive transformations:

$$a\,(b+c) = a\,b + a\,c, \qquad (a+b)\,c = a\,c + b\,c. \tag{6.31}$$

The *Collect_terms* Operator

The collection of coefficients of like terms in a polynomial occurs frequently in algebraic manipulation. During automatic simplification this operation is applied only to monomials with coefficient parts that are rational numbers. The collection of (rational and non-rational) coefficients is obtained with the *Collect_terms* operator. The goal of this operator is given in the next definition.

Definition 6.42. *An algebraic expression u is in* **collected form** *in a set S of generalized variables if it satisfies one of the following properties:*

1. *u is a GME in S.*

2. *u is a sum of GMEs in S with distinct variable parts.*

The definition is similar to Definition 6.16 for general polynomial expressions (page 225), except now property (2) requires that the variable parts be distinct.

Example 6.43. The expression

$$(2\,a + 3\,b)\,x\,y + (4\,a + 5\,b)\,x \tag{6.32}$$

is in collected form in $S = \{x, y\}$, where the two distinct variable parts are $x\,y$ and x. On the other hand, the expanded form of Expression (6.32)

$$2\,a\,x\,y + 3\,b\,x\,y + 4\,a\,x + 5\,b\,x \tag{6.33}$$

is not in collected form (in S) because there are two monomials with variable part $x\,y$ and two with variable part x.

The collected form depends, of course, on which expressions are taken as the generalized variables. For example, if $S = \{a, b\}$, the collected form of Expression (6.33) is $(2\,x\,y + 4\,x)\,a + (3\,x\,y + 5\,x)\,b$. If $S = \{a, b, c, d, x, y\}$, then Expression (6.33) is already in collected form because the four monomials in the sum have distinct variable parts.   □

Example 6.44. Strictly speaking, automatic simplification prevents the transformation of some expressions to collected form. For example, when $S = \{x\}$, the collected form of $a\,x + b\,x + c + d$ is $(a + b)\,x + (c + d)$. However, the automatic simplification rules remove the parentheses from $(c+d)$, giving two monomials c and d with variable part the integer 1. Given that our procedures operate within the context of automatic simplification, this situation is unavoidable.   □

Procedure $Collect\_terms(u, S)$;
Input
  u : an algebraic expression;
  S: a non-empty set of generalized variables;
Output
  the collected form of u or the global symbol **Undefined** if u is
  not a GPE in S;
Local Variables
  $f, combined, i, j, N, T, v$;
Begin
1   **if** $Kind(u) \neq$ " $+$ " **then**
2   **if** $Coeff_var_monomial(u, S) = $ **Undefined then**
3        $Return(\textbf{Undefined})$
4   **else** $Return(u)$
5   **else**
6   **if** $u \in S$ **then** $Return(u)$;
7      $N := 0$;
8   **for** $i := 1$ **to** $Number\_of\_operands(u)$ **do**
9        $f := Coeff\_var\_monomial(Operand(u, i), S)$;
10   **if** $f = $ **Undefined then** $Return(\textbf{Undefined})$
11   **else**
12         $j := 1$;
13         $combined := \textbf{false}$;
14   **while not** $combined$ **and** $j \leq N$ **do**
15   **if** $Operand(f, 2) = Operand(T[j], 2)$ **then**
16             $T[j] := [Operand(f, 1) + Operand(T[j], 1), Operand(f, 2)]$;
17             $combined := \textbf{true}$;
18           $j := j + 1$;
19   **if not** $combined$ **then**
20           $T[N + 1] := f$;
21           $N := N + 1$;
22     $v := 0$;
23   **for** $j := 1$ **to** N **do**
24       $v := v + Operand(T[j], 1) * Operand(T[j], 2)$;
25     $Return(v)$
End

Figure 6.9. An MPL procedure that transforms an algebraic expression to collected form. (Implementation: Maple (txt), Mathematica (txt), MuPAD (txt).)

A procedure $Collect\_terms$ that obtains a collected form is given in Figure 6.9. The procedure returns either a collected form or the symbol **Undefined** when u is not a GPE in S.

This procedure uses an array T that keeps track of the coefficients of the various monomials. Most computer algebra languages allow arrays to be used in this way. In addition, at lines 2 and 9, the procedure uses the operator $Coeff\_var\_monomial(u, S)$ which is described in Exercise 3, page 237. When u is a GME in S, this operator returns a two element list with the coefficient and variable parts of u and otherwise returns the symbol **Undefined**.

$Collect\_terms$ begins (lines 1-5) by checking if an expression u, which is not a sum, is a monomial and when this is so, returns u which is in collected form (Definition 6.42(1)). The remainder of the procedure applies to sums. At line 6 we check if the sum u is in S, which means it is in collected form. In lines 7-21, we create an array T with entries that are two operand lists that contain the coefficient and distinct variable parts obtained so far. For each operand of u, we obtain, at line 9, a list f with its coefficient and variable parts, and then check if the variable part corresponds to the variable part of some earlier operand of u that is in T (lines 12-18). If the variable part of f corresponds to an earlier variable part, the appropriate element of T is reassigned (line 16), and *combined* is assigned the symbol **true**, which terminates the **while** loop. If the variable part of f does not correspond to the variable part of some $T[j]$, for $1 \leq j \leq N$, it is added to the array T (lines 19-21). Finally, in lines 22-24, we use T to create the new expression v that is the collected form.

Observe that Definition 6.42 and $Collect\_terms$ require that u be a polynomial in S. This means, for $u = a\,x + \sin(x)\,x + b$, the operator is unable to collect coefficients in x because the expression is not even a polynomial in x. We can avoid the limitation by eliminating the free-of tests from the procedure $Coeff\_var\_monomial$ that is called in lines 2 and 10. We leave these modifications to the reader (Exercise 3).

The *Algebraic_expand* Operator

In an algebraic sense, the *Algebraic_expand* operator applies the two distributive transformations in (6.31) in a left to right fashion to products and powers that contain sums. With these transformations, the operator obtains manipulations such as:

$$(x + 2)(x + 3)(x + 4) \quad \rightarrow \quad x^3 + 9\,x^2 + 26\,x + 24, \tag{6.34}$$

$$(x + y + z)^3 \quad \rightarrow \quad x^3 + y^3 + z^3 + 3\,x^2\,y + 3\,x^2\,z + 3\,y^2\,x$$
$$+ \; 3\,y^2\,z + \; 3\,z^2\,x + 3\,z^2\,y + 6\,x\,y\,z, \tag{6.35}$$

$$(x + 1)^2 + (y + 1)^2 \quad \rightarrow \quad x^2 + 2\,x + y^2 + 2\,y + 2, \tag{6.36}$$

$$((x + 2)^2 + 3)^2 \quad \rightarrow \quad x^4 + 8\,x^3 + 30\,x^2 + 56\,x + 49. \tag{6.37}$$

The last two examples show that *Algebraic_expand* is recursive.

There are, however, other instances where it is less certain what the operator should do. For example, should *Algebraic_expand* perform the following manipulations?

$$\frac{a}{(x+1)(x+2)} \quad \rightarrow \quad \frac{a}{x^2 + 3x + 2}, \tag{6.38}$$

$$(x+y)^{3/2} \quad \rightarrow \quad x(x+y)^{1/2} + y(x+y)^{1/2}. \tag{6.39}$$

The first example differs from those above because a denominator contains a product of sums, while the second example involves non-integer exponents.

The next definition gives the form of the output of our *Algebraic_expand* operator.

Definition 6.45. *An algebraic expression u is in* **expanded form** *if the set Variables(u) does not contain a sum.*

According to this definition, the expressions on the left in (6.34)-(6.37) are in unexpanded form, while those on the right are in expanded form. For example,

$$Variables((x+2)(x+3)(x+4)) \rightarrow \{x+2, x+3, x+4\},$$

while for the expanded form of this expression,

$$Variables(x^3 + 9x^2 + 26x + 24) \rightarrow \{x\}.$$

On the other hand, the expressions on the left in (6.38) and (6.39) are already in expanded form, and so our *Algebraic_expand* operator does not obtain the manipulations shown for these expressions.

Definition 6.45 only makes sense if it is understood in the context of automatic simplification. Without this context, some expressions that are obviously not in expanded form satisfy the definition. For example, $u = \left((x+1)^2\right)^2$ is certainly not in expanded form, and since automatic simplification obtains the transformation

$$\left((x+1)^2\right)^2 \rightarrow (x+1)^4, \tag{6.40}$$

we have *Variables*$(u) = \{x+1\}$. On the other hand, without the transformation in (6.40), *Variables*$(u) = \{(x+1)^2\}$, which does not contain a sum.

The Integer Exponent Case. We describe first a simplified version of the *Algebraic_expand* algorithm which applies to algebraic expressions u with the restriction that all powers in u have integer exponents.

Procedures for expansion in this setting are given in Figures 6.10 and 6.11. The procedure *Algebraic_expand(u)* first recursively expands the operands of sums, products, and powers with positive integer exponents (lines 3, 6, and 11), and then calls on *Expand_product* and *Expand_power* to apply the distributive laws to products and powers (lines 6 and 11). Line 12 is invoked when u is not a sum, product, or power.

The procedure *Expand_product(r, s)*, which expands the product of two expanded expressions, uses a recursive approach to apply the right and left distributive laws. If r is a sum, it applies the right distributive law (line 3), and if s is a sum, it apples the left distributive law by a recursive call with the operands interchanged (line 5). (If both r and s are sums, both distributive laws are applied through recursion.) Line 7, which serves as a termination condition for the recursion, applies when neither r nor s is a sum. The assumption that all exponents are integers is essential here because without it the output of the procedure may not be in expanded form (see Expression (6.43) below).

The procedure *Expand_power(u, n)*, which expands an expanded expression u to an integer power $n \geq 2$, is given in Figure 6.11. When u is a sum, the expanded form is obtained by letting

$$f = Operand(u, 1), \quad r = u - f,$$

and applying the binomial expansion

$$u^n = (f + r)^n = \sum_{k=0}^{n} \left(\left(\frac{n!}{k!(n-k)!} f^{n-k} \right) r^k \right).$$

Observe that the automatically simplified form of the expression

$$\frac{n!}{k!(n-k)!} f^{n-k}$$

is in expanded form. Indeed, f^{n-k} is in expanded form because f is in expanded form (by recursion), and f is not a sum because it is the operand of a sum. The assumption that all exponents are integers is used here because without it f^{n-k} may not be in expanded form (see Expression (6.45) below). On the other hand, the base of r^k can be a sum, and so this power must be expanded recursively. This operation is performed in line 7, where *Expand_product* is used to expand the product of these two expressions.

Procedure *Algebraic_expand(u)*;
Input
    u : an algebraic expression where all exponents of powers are integers;
Output
    the expanded form of *u*;
Local Variables
    v, base, exponent;
Begin
1   **if** *Kind(u)* = " + " **then**
2   *v* := *Operand(u, 1)*;
3   *Return(Algebraic_expand(v) + Algebraic_expand(u − v))*
4   **elseif** *Kind(u)* = " ∗ " **then**
5   *v* := *Operand(u, 1)*;
6   *Return(Expand_product(Algebraic_expand(v), Algebraic_expand(u/v)))*
7   **elseif** *Kind(u)* = " ∧ " **then**
8   *base* := *Operand(u, 1)*;
9   *exponent* := *Operand(u, 2)*;
10   **if** *Kind(exponent)* = **integer and** *exponent* ≥ 2 **then**
11   *Return(Expand_power(Algebraic_expand(base), exponent))*;
12   *Return(u)*
End

Procedure *Expand_product(r, s)*;
Input
    r,s : expanded algebraic expressions, where all exponents of powers are
        integers;
Output
    the expanded form of *r* ∗ *s*;
Local Variables
    f;
Begin
1   **if** *Kind(r)* = " + " **then**
2   *f* := *Operand(r, 1)*;
3   *Return(Expand_product(f, s) + Expand_product(r − f, s))*;
4   **elseif** *Kind(s)* = " + " **then**
5   *Return(Expand_product(s, r))*
6   **else**
7   *Return(r ∗ s)*
End

Figure 6.10. MPL procedures for *Algebraic_expand* and *Expand_product*. (Implementation: Maple (txt), Mathematica (txt), MuPAD (txt).)

Procedure *Expand_power(u, n)*;
Input
    u : an expanded algebraic expression where all exponents of powers are
        integers;
    n : a non-negative integer;
Output
    the expanded form of u^n;
Local Variables
    f, r, k, s, c;
Begin
1   **if** *Kind(u)* = " + " **then**
2        $f := Operand(u, 1)$;
3        $r := u - f$;
4        $s := 0$;
5   **for** $k := 0$ **to** n **do**
6            $c := n!/(k! * (n - k)!)$;
7            $s := s + Expand\_product(c * f^{n-k}, Expand\_power(r, k))$;
8   *Return(s)*
9   **else**
10       $Return(u^n)$
End

Figure 6.11. An MPL procedure for *Expand_power*. (Implementation: Maple
(txt), Mathematica (txt), MuPAD (txt).)

The Non-integer Exponent Case. If u contains powers with non-integer exponents, the *Algebraic_expand* operator may return an expression that is not in expanded form. To see how this happens, let's suppose that the transformations

$$u^v\, u^w \;\; = \;\; u^{v+w}, \tag{6.41}$$

$$(u^v)^n \;\; = \;\; u^{n\,v}, \quad n \text{ an integer}, \tag{6.42}$$

are applied from left to right during automatic simplification[6]. (The two transformations hold for both the real and complex interpretations of the

---

[6] In Maple, automatic simplification obtains the transformation (6.41), when v and w are rational numbers, and (6.42), when v is a rational number.

In Mathematica, automatic simplification obtains the transformations (6.41) and (6.42).

In MuPAD, automatic simplification obtains the transformation (6.41), when v and w are rational numbers, and (6.42).

For a summary of the power transformations in Maple, Mathematica, and MuPAD, see Cohen [24], Section 3.1.

power operation.) In this context, *Algebraic_expand* together with the transformation (6.41) obtains

$$Algebraic\_expand\left((x\,(y+1)^{3/2}+1)\,(x\,(y+1)^{3/2}-1)\right)$$
$$\rightarrow \quad x^2\,(y+1)^3-1 \tag{6.43}$$
$$= \quad x^2\,y^3+3\,x^2\,y^2+3\,x^2\,y+x^2-1, \tag{6.44}$$

and together with the transformation (6.42) obtains

$$Algebraic\_expand\left((x\,(y+1)^{1/2}+1)^4\right)$$
$$\rightarrow \quad x^4\,(y+1)^2+4\,x^3(y+1)^{3/2} \tag{6.45}$$
$$+6\,x^2(y+1)+4\,x\,(y+1)^{1/2}+1$$
$$= \quad x^4\,y^2+2\,x^4\,y+x^4+4\,x^3\,(y+1)^{3/2} \tag{6.46}$$
$$+6\,x^2\,y+6\,x^2+4\,x\,(y+1)^{1/2}+1.$$

Algebraic_expand obtains Expressions (6.43) and (6.45), and the expanded forms are shown in Expressions (6.44) and (6.46). In both examples, the output of *Algebraic_expand* is not in expanded form, because in each case the output of the *Variables* operator contains the generalized variable $y+1$. In the first example, the input to *Algebraic_expand* is a product, and the output contains a new product and a new power that are not in expanded form. This situation arises from line 7 in *Expand_product* when $r=s=x\,(y+1)^{3/2}$, and so by the transformation (6.41), $r*s=x^2\,(y+1)^3$ which is not in expanded form. In the second example, the input is a power, and the output contains new products and powers that are not in expanded form. This situation arises from line 7 in *Expand_power*. For example, when $f=x\,(y+1)^{1/2}$ and $n-k=4$, the transformation (6.42) implies $f^{n-k}=x^4\,(y+1)^2$ which is not in expanded form.

One way to expand an expression u_0 with non-integer exponents is to apply a sequence of expansions

$$u_1 \quad := \quad Algebraic\_expand(u_0),$$
$$u_2 \quad := \quad Algebraic\_expand(u_1),$$
$$\vdots$$
$$u_i \quad := \quad Algebraic\_expand(u_{i-1}),$$
$$\vdots$$

where the process stops when $u_i=u_{i-1}$. The problem with this approach is that it performs unnecessary work by trying to expand parts of an expression that are already in expanded form. Another approach is to modify the algorithm in Figures 6.10 and 6.11 so that new unexpanded products and powers obtained with *Expand_product* and *Expand_power* are expanded, although we must take care to avoid introducing redundant recursion or infinite recursive loops. We leave the details of this modification to the reader (Exercise 9(a)).

Extensions of the *Algebraic_expand* Operator. There are two extensions of the *Algebraic_expand* operator that obtain manipulations beyond the expanded form described in Definition 6.45. The first extension is based on the **expand** operator in the Macsyma system which returns an expression with the following properties.

1. Each complete sub-expression of an expression is in expanded form,

2. The denominator of each complete sub-expression in an expression is in expanded form.

These properties include Definition 6.45 and imply the following expansions:

$$\sin(a\,(b+c)) \quad \rightarrow \quad \sin(a\,b + a\,c), \tag{6.47}$$

$$\frac{a}{b\,(c+d)} \quad \rightarrow \quad \frac{a}{b\,c + b\,d}. \tag{6.48}$$

In each of these examples, the expression on the left satisfies Definition 6.45 because

$$Variables(\sin(a\,(b+c))) \quad \rightarrow \quad \{\sin(a\,(b+c))\},$$

$$Variables\left(\frac{a}{b\,c + b\,d}\right) \quad \rightarrow \quad \{a, 1/b, 1/(c+d)\}.$$

However, in Expression (6.47), $\sin(a\,(b+c))$ has a complete sub-expression that is not in expanded form, and in Expression (6.48), the denominator of

$$\frac{a}{b\,(c+d)}$$

is not in expanded form. An extension of the *Algebraic_expand* operator that obtains these transformations is described in Exercise 8.

The second extension of *Algebraic_expand*, which has to do with fractional exponents, is based on the **Expand** operator in Mathematica. Let u be an algebraic expression, and let f be a positive fraction (that is not an integer). According to Definition 6.45, the expression u^f is in expanded form. However, another expanded form for u^f is obtained by separating f

into the sum of an integer and a fraction m with $0 < m < 1$. We obtain this representation with

$$f = \lfloor f \rfloor + m,$$

where

$$\lfloor f \rfloor = \text{largest integer} \leq f, \quad m = f - \lfloor f \rfloor.$$

The function $\lfloor f \rfloor$ is called the *floor function* of f. We have

$$u^f = u^m \, u^{\lfloor f \rfloor}, \tag{6.49}$$

and obtain an expanded form by expanding $u^{\lfloor f \rfloor}$ and multiplying each term of this expansion by u^m. For example, with $\lfloor 5/2 \rfloor = 2$, we have

$$\begin{aligned}
(x+1)^{5/2} &= (x+1)^{1/2}(x+1)^2 \\
&= (x+1)^{1/2} x^2 + 2(x+1)^{1/2} x + (x+1)^{1/2}. \tag{6.50}
\end{aligned}$$

An extension of *Expand_power* that obtains this transformation is described in Exercise 9(b).

Exercises

1. Suppose u is in collected form with respect to a set S.

   (a) Is u also in collected form with respect to a subset of S?

   (b) Is each complete sub-expression of u also in collected form with respect to S?

2. Let u be a sum. A common algebraic operation is to combine terms that have denominators with the same variable part into a single term. For example,

$$\frac{a}{2bc} + \frac{d}{3bc} = \frac{a/2 + d/3}{bc}. \tag{6.51}$$

   In this example, both denominators have the variable part bc. Give a procedure *Combine(u)* that combines terms in a sum u whose (non-constant) denominators differ by at most a rational number factor into a single term. If u is not a sum, then return u. This exercise requires a *Denominator* operator. Most CAS languages have this operator (see Figure 4.1 on page 124), and transformation rules for the operator are given in Section 6.5. *Hint:* This operation can be performed by the *Collect_terms(u, S)* procedure with the appropriate generalized variables in S.

3. Let u be an algebraic expression, and let S be a set of generalized variables. Give a procedure *Collect_terms_2(u, S)* that collects coefficients in S but doesn't require that u be a GPE in S. For example,

$$Collect\_terms\_2(a\,x + \sin(x)\,x + b, \{x\}) \to (a + \sin(x))\,x + b.$$

4. Explore the capacity of the algebraic expand operator in a CAS. How does it compare with the version of the operator described in this section? (Use **expand** in Maple and MuPAD, and **Expand** in Mathematica.)

5. Let u be an algebraic expression. The operator $Distribute(u)$, which provides a fast way to apply the distributive transformation, can replace $Algebraic\_expand$ in some situations. It is defined using the following rules.

   (a) If u is not a product, then

   $$Distribute(u) \to u.$$

   (b) If u is a product, then

      i. If u does not have an operand that is a sum, then

      $$Distribute(u) \to u.$$

      ii. Suppose u has an operand that is a sum, and let v be the first such operand. Form a new sum by multiplying the remaining operands of u by each operand of v. Return this sum.

   For example,

   $$Distribute\left(a\,(b+c)\,(d+e)\right) \to a\,b\,(d+e) + a\,c\,(d+e),$$

   $$Distribute\left(\frac{x+y}{x\,y}\right) \to 1/y + 1/x.$$

   Give a procedure for $Distribute(u)$.

6. Give a procedure $Expand\_main\_op(u)$ that expands only with respect to the main operator of u. In other words, the operator does not recursively expand the operands of sums, products, or powers before it applies the distributive transformations. For example,

   $$Expand\_main\_op\left(x\,\left(2+(1+x)^2\right)\right) \quad \to \quad 2\,x + x\,(1+x)^2,$$

   $$Expand\_main\_op\left(\left(x+(1+x)^2\right)^2\right) \quad \to \quad x^2 + 2\,x\,(1+x)^2 + (1+x)^4.$$

7. Let T be a set of expressions that are sums. Give procedures for an operator $Expand\_restricted(u, T)$ which applies the distributive laws as $Algebraic\_expand$ does, except that it does not apply the laws to members of T. For example, for $u = (x+a)^2\,(x+b)$,

   $$Expand\_restricted(u, \{x+a\}) \to (x+a)^2 x + (x+a)^2\,b,$$

   $$Expand\_restricted(u, \{x+b\}) \to (x+b)\,x^2 + 2\,a\,(x+b)\,x + a^2(x+b),$$

   $$Expand\_restricted(u, \{x+a, x+b\}) \to (x+a)^2(x+b).$$

8. Let u be an algebraic expression. Modify the expand algorithm so that it returns an expression with properties (1) and (2) on page 256. Your procedure should obtain the expansions in (6.38), (6.47), and (6.48). Since these properties require the expansion of denominators, there is the possibility that a denominator expands and simplifies to 0. For example, this occurs with

$$\frac{1}{x^2 + 1 - x\,(x+1)}.$$

Make sure your procedures check for this situation and return the global symbol **Undefined** when it occurs. This exercise requires the *Numerator* and *Denominator* operators. Most CAS languages have these operators (see Figure 4.1 on 124), and transformation rules for these operators are given in Section 6.5.

9. (a) Modify the *Algebraic_expand* algorithm so that it obtains the expanded form when the input expressions include powers with fraction exponents. Assume that the transformations in (6.41) and (6.42) are included in automatic simplification.

   (b) Modify the *Expand_power* procedure in part (a) so that it also obtains the expansions using the decomposition of fraction powers in (6.49) and (6.50). Most computer algebra languages have an operator to compute $\lfloor N \rfloor$. (In Maple and MuPAD use `floor`, and in Mathematica use `Floor`.)

6.5   General Rational Expressions

In a mathematical sense, a *rational expression* is defined as a quotient of two polynomials. In this section we discuss the rational expression structure of an algebraic expression and describe an algorithm that transforms an expression to a particular rational form.

Definition 6.46. **(Mathematical Definition)** *Let $S = \{x_1, \ldots, x_m\}$ be a set of generalized variables. An algebraic expression u is a* **general rational expression** *(GRE) in S if it has the form $u = p/q$, where p and q are GPEs in S.*

Example 6.47.

$$\frac{x^2 - x + y}{x + 4}, \qquad S = \{x\},$$

$$\frac{x^2 \sin(y) - x \sin^2(y) + 2\,(z+1)}{x + \sin(y)}, \qquad S = \{x, \sin(y)\},$$

$$x^2 + b\,x + c, \qquad S = \{x\}.$$

For each example, we have given one possible choice for S. Notice that the definition is interpreted in a broad sense to include GPEs for which the denominator is understood to be 1. □

The *Numerator* and *Denominator* Operators. To determine if an expression is a GRE, we must define precisely the numerator and denominator of the expression. The *Numerator* and *Denominator* operators, which are used for this purpose, are defined by the following transformation rules.

Definition 6.48. *Let u be an algebraic expression.*

ND-1. *If u is a fraction, then*

$$Numerator(u) \rightarrow Operand(u, 1),$$
$$Denominator(u) \rightarrow Operand(u, 2).$$

ND-2. *Suppose u is a power. If the exponent of u is a negative integer or a negative fraction, then*

$$Numerator(u) \rightarrow 1, \quad Denominator(u) \rightarrow u^{-1},$$

otherwise

$$Numerator(u) \rightarrow u, \quad Denominator(u) \rightarrow 1.$$

ND-3. *Suppose u is a product and $v = Operand(u, 1)$. Then*

$$Numerator(u) \rightarrow Numerator(v) * Numerator(u/v),$$

$$Denominator(u) \rightarrow Denominator(v) * Denominator(u/v).$$

ND-4. *If u does not satisfy any of the previous rules, then*

$$Numerator(u) \rightarrow u, \quad Denominator(u) \rightarrow 1.$$

Example 6.49. Consider the expression $u = (2/3) \dfrac{x(x+1)}{x+2} y^n$. Then

$$Numerator(u) \rightarrow 2x(x+1)y^n, \quad Denominator(u) \rightarrow 3(x+2). \quad □$$

The *Numerator* and *Denominator* operators are defined in terms of the tree structure of an expression and are interpreted in the context of automatic simplification. Although the operators are adequate for our purposes, the next two examples show in some cases they give unusual results.

Example 6.50. Consider the expression

$$\frac{1}{x} + \frac{1}{y}.$$

Certainly, if we transform the expression to

$$\frac{x + y}{x\,y},$$

it is clear which expression is the numerator and which is the denominator. The definition, however, does not include this transformation as part of the simplification context, and so the numerator is

$$\frac{1}{x} + \frac{1}{y}$$

and the denominator is 1.   □

Example 6.51. Consider the expression $x^{-r^2 - 4\,r - 5}$. In this case, the exponent is negative for all real values of r. However, since the exponent of the expression is not a negative integer or fraction, the numerator is $x^{-r^2 - 4\,r - 5}$ and the denominator is 1.   □

Modifications of the *Numerator* and *Denominator* operators that address the issues in the last two examples are described in Exercise 4.

We give next a definition of a general rational expression that is more suitable for computational purposes.

Definition 6.52. (**Computational Definition**) *Let* $S = \{x_1, \ldots, x_m\}$ *be a set of generalized variables. An algebraic expression* u *is a* **general rational expression** *(GRE) in* S *if Numerator(u) and Denominator(u) are GPEs in* S.

The *Rational_gre* Operator

Definition 6.53. *Let* u *be an algebraic expression, and let* v *be either a generalized variable* x *or a set* S *of generalized variables. The operator*

$$Rational\_gre(u, v)$$

returns **true** *whenever* u *is a GRE in* $\{x\}$ *or* S *and otherwise returns* **false**. *The operator is defined by the following transformation rule:*

$Rational\_gre(u, v) \quad \rightarrow$
$Polynomial\_gpe(Numerator(u), v)$ **and** $Polynomial\_gpe(Denominator(u), v)$

where the Polynomial_gpe operator is given in Figure 6.6 on page 228.

Example 6.54.

$$Rational\_gre\left(\frac{x^2+1}{2\,x+3},\ x\right) \quad \rightarrow \quad \textbf{true},$$

$$Rational\_gre\left(\frac{1}{x}+\frac{1}{y},\ \{x,y\}\right) \quad \rightarrow \quad \textbf{false}. \tag{6.52}$$

$\square$

The *Rational_variables* Operator.   The *Rational_variables* operator defines a natural set of generalized variables for a rational expression.

Definition 6.55. *Let u be an algebraic expression. The operator*

$$Rational\_variables(u)$$

is defined by the transformation rule:

$Rational\_variables(u) \quad \rightarrow$
$\qquad Variables(Numerator(u)) \quad \cup \quad Variables(Denominator(u)),$

where the Variables operator is given in Definition 6.19 on page 227.

Example 6.56.

$$Rational\_variables\left(\frac{2\,x+3\,y}{z+4}\right) \quad \rightarrow \quad \{x,y,z\},$$

$$Rational\_variables\left(\frac{1}{x}+\frac{1}{y}\right) \quad \rightarrow \quad \left\{\frac{1}{x},\frac{1}{y}\right\}.$$

There is a natural way to view $(2\,x+3\,y)/(z+4)$ as a GRE in x, y, and z. On the other hand, $1/x + 1/y$ is not a GRE in x and y (see Expression (6.52)) but can be viewed as a GRE in the two generalized variables $1/x$ and $1/y$. $\square$

Rationalization of Algebraic Expressions

The rationalization process, which is based on the transformation that combines operands in a sum over a common denominator, transforms an algebraic expression to a form with a more appropriate set of generalized variables. When the process is applied (in a recursive manner), it obtains the following transformations:

$$\frac{a}{b} + \frac{c}{d} \to \frac{a\,d + b\,c}{b\,d},$$

$$1 + \frac{1}{1 + 1/x} \to \frac{2\,x + 1}{x + 1},$$

$$\frac{1}{\left(1 + \dfrac{1}{x}\right)^{1/2}} + \left(1 + \frac{1}{x}\right)^{3/2} \to \frac{x^2 + (x+1)^2}{x^2 \left(\dfrac{x+1}{x}\right)^{1/2}}. \tag{6.53}$$

The goal of rationalization is described in the following definition.

Definition 6.57. *An algebraic expression u is in* **rationalized form** *if it satisfies one of the following properties:*

1. *u is an integer, fraction, symbol, factorial, or function form.*

2. *u is any other type, and consider u as a rational expression in*

$$S = Rational\_variables(u).$$

Then,

(a) each expression v in S is in rationalized form with

$$Denominator(v) = 1,$$

(b) the coefficient part of each of the monomials in $Numerator(u)$ and $Denominator(u)$ is an integer.

Observe that Rule 2(a) is recursive. As usual, we interpret this definition in the context of automatic simplification.

Some examples will help clarify the definition.

Example 6.58. The expression $a/b + c/d$ is not in rationalized form because

$$Rational\_variables(a/b + c/d) \to \{a, 1/b, c, 1/d, \},$$

and so property 2(a) of Definition 6.57 is not satisfied. However, $(a\,d + b\,c)/(b\,d)$ is in rationalized form because

$$Rational\_variables\left(\frac{a\,d + b\,c}{b\,d}\right) \to \{a, b, c, d\},$$

and the coefficient part of each of the monomials $a\,d$, $b\,c$, and $b\,d$ is 1.

The expression

$$1 + \frac{1}{1 + 1/x}$$

is not in rationalized form because

$$Rational\_variables\left(1 + \frac{1}{1 + 1/x}\right) \to \left\{\frac{1}{1 + 1/x}\right\},$$

and so property 2(a) of Definition 6.57 is not satisfied. This expression can be transformed to

$$\frac{2\,x + 1}{x + 1},$$

which is in rationalized form because

$$Rational\_variables\left(\frac{2\,x + 1}{x + 1}\right) \to \{x\},$$

and the coefficient parts of all monomials in the numerator and denominator are integers.

The expression $a + b/2$ is not in rationalized form because the coefficient part of $b/2$ is not an integer, and so property 2(b) in Definition 6.57 is not satisfied. However, its sum $(2\,a + b)/2$ is in rationalized form. □

The *Rationalize_expression* Operator.   The operator

$$Rationalize\_expression(u)$$

n transforms an algebraic expression u to an equivalent expression in rationalized form. The operator is understood to operate in an automatic simplification context that includes the power transformations[7]

---

[7] In Maple, automatic simplification obtains the transformation (6.54) when v and w are rational numbers, (6.55) when v is a rational number, and (6.56).

In Mathematica, automatic simplification obtains the transformation (6.54), (6.55), and (6.56).

In MuPAD, automatic simplification obtains the transformation (6.54) when v and w are rational numbers, (6.55), and (6.56).

For a summary of power transformation rules in Maple, Mathematica, and MuPAD, see Cohen [24], Section 3.1.

Procedure $Rationalize\_expression(u)$;
Input
 u : an algebraic expression;
Output
 a rationalized form of u;
Local Variables f, g, r;
Begin
1 **if** $Kind(u) = " \wedge "$ **then**
2 $Return\left(Rationalize\_expression(Operand(u,1))^{Operand(u,2)}\right)$
3 **elseif** $Kind(u) = " * "$ **then**
4 $f := Operand(u,1)$;
5 $Return(Rationalize\_expression(f) * Rationalize\_expression(u/f))$
6 **elseif** $Kind(u) = " + "$ **then**
7 $f := Operand(u,1)$;
8 $g := Rationalize\_expression(f)$;
9 $r := Rationalize\_expression(u - f)$;
10 $Return(Rationalize\_sum(g,r))$
11 **else**
12 $Return(u)$
End

Procedure $Rationalize\_sum(u, v)$;
Input
 u, v : algebraic expressions in rationalized form;
Output
 an algebraic expression in rationalized form;
Local Variables m, n, r, s;
Begin
1 $m := Numerator(u)$;
2 $r := Denominator(u)$;
3 $n := Numerator(v)$;
4 $s := Denominator(v)$;
5 **if** $r = 1$ **and** $s = 1$ **then**
6 $Return(u + v)$
7 **else**
8 $Return(Rationalize\_sum(m * s,\ n * r)/(r * s))$
End

Figure 6.12. An MPL algorithm that rationalizes an algebraic expression. (Implementation: Maple (txt), Mathematica (txt), MuPAD (txt).)

$$u^v u^w \;\rightarrow\; u^{v+w}, \tag{6.54}$$

$$(u^v)^n \;\rightarrow\; u^{v\,n}, \tag{6.55}$$

$$(u\,v)^n \;\rightarrow\; u^n v^n, \tag{6.56}$$

where u, v, and w are algebraic expressions and n is an integer. These transformations hold for both the real and complex interpretations of the power operation.

Procedures that transform an expression to rationalized form are given in Figure 6.12. In the main procedure *Rationalize_expression*, in lines 1-2 a power is rationalized by recursively rationalizing its base. For example,

$$Rationalize\_expression\left((1+1/x)^2\right) \;\rightarrow\; \frac{(x+1)^2}{x^2},$$

where the transformation is obtained by rationalizing the base $1+1/x$ and then using the transformation (6.56). Notice that we rationalize the base even when the exponent is not an integer because the base may appear with an integer exponent later in the computation as a result of the rationalization process (see Example 6.59 below). Unfortunately, this means that some expressions that are already in rationalized form are transformed to another rationalized form. For example,

$$Rationalize\_expression\left((1+1/x)^{1/2}\right) \;\rightarrow\; \left(\frac{x+1}{x}\right)^{1/2}.$$

In lines 3-5, a product is rationalized by recursively rationalizing each of its operands.

In lines 6-10, a sum is rationalized by first rationalizing its operands and then combining the operands over a common denominator (line 10). The actual sum transformation occurs in the *Rationalize_sum* procedure that performs the transformation

$$m/r + n/s \;\rightarrow\; \frac{m\,s + n\,r}{r\,s}. \tag{6.57}$$

Notice that *Rationalize_sum* is recursive (line 8) because the sum in the numerator of the right side of (6.57) may not be in rationalized form (see Example 6.59 below). The termination condition for the recursion is in lines 5-6. We have separated the computation into two procedures to avoid some redundant recursion.

Example 6.59. In this example we outline the steps in a rationalization that requires both the rationalization of powers with non-integer exponents and the recursive step in *Rationalize_sum*. Consider the expression[8]

---

[8] For clarity, we use notation with the quotient operator even though quotients are transformed to products and powers by automatic simplification.

$$\frac{1}{\left(1+\dfrac{1}{x}\right)^{1/2}} + \left(1+\frac{1}{x}\right)^{3/2}.$$

Rationalizing the two operands of the sum, we obtain

$$\frac{1}{\left(\dfrac{x+1}{x}\right)^{1/2}} + \left(\frac{x+1}{x}\right)^{3/2}.$$

Applying the sum transformation in (6.57) followed by the power transformations in (6.54) and (6.55), we obtain

$$\frac{1+\dfrac{(x+1)^2}{x^2}}{\left(\dfrac{x+1}{x}\right)^{1/2}},$$

where the sum in the numerator is not in rationalized form. Again applying the transformation (6.57) to the numerator, we obtain with automatic simplification

$$\frac{x^2+(x+1)^2}{x^2\left(\dfrac{x+1}{x}\right)^{1/2}},$$

which is in rationalized form. $\qquad\qquad\qquad\qquad\qquad\qquad\qquad\square$

Example 6.60. In order for the algorithm to obtain a rationalized form, distributive transformations that undo a rationalization cannot be included in automatic simplification. A problem arises with both the Maple and MuPAD systems in which integers and fractions are automatically distributed over sums. For example, in Maple or MuPAD, implementations of the algorithm attempt to transform $a+b/2$ to $(2a+b)/2$, but then automatic simplification transforms it back to $a+b/2$. $\qquad\square$

Rational-Expanded Form

Since algebraic expansion is not part of the simplification context of rationalization, the *Rationalize_expression* operator may return an expression with the numerator or denominator in unexpanded form. For example,

$$Rationalize\_expression(a/b + c/d + e/f) \rightarrow \frac{adf + b(cf + de)}{bdf}.$$

The following definition combines rationalization and expansion.

Definition 6.61. *An algebraic expression* u *is in* **rational-expanded** *form if it satisfies the following two properties:*

1. u *is in rationalized form.*

2. *Numerator(u) and Denominator(u) are in algebraic expanded form.*

The next example shows that there is an involved interaction between the rationalization and expansion operations.

Example 6.62. Consider the expression

$$\frac{\left(\sqrt{\frac{1}{(x+y)^2+1}}+1\right)\left(\sqrt{\frac{1}{(x+y)^2+1}}-1\right)}{x+1}.$$

This expression is in rationalized form, but not rational-expanded form. Expanding the numerator we obtain

$$\frac{\frac{1}{(x+y)^2+1}-1}{x+1},$$

which is not in rationalized form. Transforming this expression to rationalized form we obtain

$$\frac{-(x+y)^2}{\left((x+y)^2+1\right)(x+1)},$$

which again is not in rational-expanded form. Expanding the numerator and denominator, we obtain

$$\frac{-x^2-2\,x\,y-y^2}{x^3+x^2+2\,x^2\,y+2\,x\,y+x\,y^2+y^2+x+1},$$

which is in rational-expanded form.   □

The operator *Rational_expand(u)* transforms an algebraic expression u to rational-expanded form. The procedure for this operator is left to the reader (Exercise 3).

Normal Simplification Operators

Let **M** represent the set of algebraic expressions that do not contain factorials, function forms, or powers with non-integer exponents. For this class of expressions, the *Rational_expand* operator together with automatic simplification can always determine if an expression simplifies to 0. An operator with this property is called a *normal simplification operator* or a *zero equivalence operator* for the class **M**. For example,

$$Rational\_expand \left(\frac{1}{1/a + c/(ab)} + \frac{abc + ac^2}{(b+c)^2} - a \right) \to 0.$$

On the other hand, rationalization alone does not obtain this transformation:

$$Rationalize\_expression \left(\frac{1}{1/a + c/(ab)} + \frac{abc + ac^2}{(b+c)^2} - a \right)$$

$$\to \frac{(b+c)^2 a^2 b + \left(abc + ac^2 - a(b+c)^2 \right)(ab + ca)}{(ab+ca)(b+c)^2}.$$

Rational Simplification

Although the *Rationalize_expression* operator transforms an expression to rationalized form, it often introduces extraneous common factors into the numerator and denominator. For example, the operator obtains

$$Rationalize\_expression(x/z + y/z^2) \to \frac{z^2 x + zy}{z^3},$$

where an extraneous common factor z appears in the numerator and denominator. Although it is possible to modify the algorithm to avoid this, it is better to eliminate the common factors after rationalization because other common factors can be eliminated then as well. In Exercise 6, we describe an operator that eliminates the explicit common factors that arise during rationalization as well as some other explicit common factors.

The more interesting problem, however, involves the elimination of common factors that are implicit or hidden. For example, it is not so obvious that the expression

$$\frac{2a^3 + 22ab + 6a^2 + 7a + 6ba^2 + 12b^2 + 21b}{7a^2 - 5ab^2 - 2ba^2 - 5a + 21ab + 3b^3 - 15b} \tag{6.58}$$

has a common factor $a + 3b$ in the numerator and denominator and can be simplified to

$$\frac{2a^2 + 4b + 6a + 7}{7a - 2ab - 5 + b^2}. \tag{6.59}$$

The process of eliminating explicit and implicit common factors from the numerator and denominator of a rational expression is called *rational simplification*. One way to obtain this simplification is by factoring the numerator and denominator and cancelling the common factors. Since factorization is a time consuming process, this is usually done instead with a greatest common divisor algorithm. Since the topic is beyond the scope of this chapter, the reader may consult Cohen [24], Sections 4.2 and 6.3, for more information on this problem.

Since rational simplification is an important aspect of simplification, most computer algebra systems have some capability to perform this operation (Exercise 5).

Exercises

1. Explore the capacity of the numerator and denominator operators in a CAS. What is the simplification context of these operators in the CAS? Are the operators defined with the same transformations as the ones given in the text? (See Figure 4.1 on page 124.)

2. Let u be an algebraic expression. Give procedures for each of the following operators:

    (a) *Numerator*(u) (Definition 6.48).

    (b) *Denominator*(u) (Definition 6.48).

    (c) *Rational_gre*(u, v) (Definition 6.53).

    (d) *Rational_variables*(u) (Definition 6.55).

3. Let u be an algebraic expression. Give a procedure *Rational_expand*(u) that transforms u to rational-expanded form (Definition 6.61). Since rational expansion includes the expansion of denominators, there is the possibility that a denominator expands and simplifies to 0. For example, this occurs with

$$\frac{1}{x^2 + x - x\,(x + 1)}.$$

Make sure your procedure checks for this situation and returns the global symbol **Undefined** when it occurs. Your procedure should obtain the rational expansion in Example 6.59.

4. In this exercise we describe two modifications of the operators *Numerator* and *Denominator*.

    (a) Give procedures for *Numerator*(u) and *Denominator*(u) that rationalize u before obtaining the numerator and denominator.

    (b) Let u be an algebraic expression, and let L be a list of distinct symbols. In addition, suppose the exponent of each power in u is a multivariate polynomial in the variables in L. A modification of the definition for the *Numerator* and *Denominator* operators is obtained by determining the sign of the exponent of a power using the *Polynomial_sign* operator described in Exercise 7 on page 246. In this case, Rule ND-2 is replaced by the following rule.

ND-2. Suppose that u is a power, and let $z = Operand(u, 2)$. If

$$Polynomial\_sign(z, L) < 0,$$

then
$$Numerator(u, L) \rightarrow 1,$$

and

$$Denominator(u, L) \rightarrow$$
$$Operand(u, 1) \wedge Algebraic\_expand(-1 * z),$$

otherwise

$$Numerator(u, L) \rightarrow u, \qquad Denominator(u, L) \rightarrow 1.$$

Notice that the list L appears as an input parameter because the polynomial sign depends on the order of the symbols in L. Give procedures $Numerator(u, L)$ and $Denominator(u, L)$ that obtain the numerator and denominator of u with this modification to the ND rules.

5. Explore the rational simplification capability of a CAS. For example can the rational simplification operator in a CAS simplify Expression (6.58) to Expression (6.59)? How about the transformation

$$\frac{x^3 + \left(\sqrt{2} + \sqrt{3}\right) x^2 + \left(2\sqrt{2}\sqrt{3} - 5\right) x + \sqrt{2} - \sqrt{3}}{x^3 + \left(-\sqrt{2} + \sqrt{3}\right) x^2 + \left(-5 - 2\sqrt{2}\sqrt{3}\right) x - \sqrt{2} - \sqrt{3}}$$

$$\rightarrow \frac{x + \sqrt{2} - \sqrt{3}}{x - \sqrt{2} - \sqrt{3}},$$

which is more involved because it includes radical expressions? (See Figure 4.1 on page 124.)

6. Let u be an algebraic expression in rationalized form. In this exercise we outline an algorithm for an operator $Cancel(u)$ that performs a limited version of rational simplification. The operator can eliminate extraneous common factors introduced by the *Rationalize_expression* operator as well as some other explicit common factors. The cancellation is obtained through automatic simplification after performing a limited version of factorization on the numerator and denominator of u. The operator is based on the following operators:

(a) Let u and v be algebraic expressions. The operator

$$Common\_factors(u, v)$$

finds some factors that are common to u and v. It is defined using the following transformation rules.

CF-1. If u and v are integers then $Common\_factors(u, v)$ returns the greatest (positive) common divisor of u and v. Most computer algebra systems have an operator that obtains the greatest common divisor of integers (see Figure 4.1 on page 124).

CF-2. If u is a product, let

$$f = Operand(u, 1), \quad r = Common\_factors(f, v).$$

Then

$$Common\_factors(u, v) \to r * Common\_factors(u/f, v/r).$$

CF-3. If v is a product then

$$Common\_factors(u, v) \to Common\_factors(v, u).$$

CF-4. If none of the previous rules apply, then define

$$\text{base}(u) = \begin{cases} Operand(u, 1) & \text{if } Kind(u) = \text{"} \wedge \text{"}, \\ u & \text{otherwise}, \end{cases}$$

$$\text{exponent}(u) = \begin{cases} Operand(u, 2) & \text{if } Kind(u) = \text{"} \wedge \text{"}, \\ 1 & \text{otherwise}. \end{cases}$$

If $\text{base}(u) = \text{base}(v)$ and both $\text{exponent}(u)$ and $\text{exponent}(v)$ are positive rational numbers, then

$$Common\_factors(u, v) \to$$
$$\text{base}(u)^{Min(\{\text{exponent}(u), \text{exponent}(v)\})},$$

otherwise $Common\_factors(u, v) \to 1$.

For example, the operator obtains

$$Common\_factors\left(6\, x\, y^3,\ 2\, x^2\, y\, z\right) \to 2\, x\, y,$$

$$Common\_factors(x + y,\ a\,(x + y)) \to x + y.$$

Give a procedure for this operator.

(b) Let u be an algebraic expression. The operator $Factor\_out(u)$ performs a limited version of factorization. It is defined using the following transformation rules.

FO-1. If u is a product then $Factor\_out(u) \to Map(Factor\_out, u)$.

FO-2. If u is a power then

$$Factor\_out(u) \to Factor\_out(Operand(u,1))^{Operand(u,2)}.$$

FO-3. Suppose that u is a sum with n operands, and let

$$s = Map(Factor\_out, u).$$

If s is not a sum, then

$$Factor\_out(u) \to s.$$

Otherwise, suppose that s is a sum with operands $s_1, \ldots, s_n$, and let c be the common factor of all the s_i obtained using the *Common_factors* operator described in part (a). Then

$$Factor\_out(u) \to c\,(s_1/c + \cdots + s_n/c),$$

where the divisions are obtained with automatic simplification.

FO-4. If none of the previous rules apply, then $Factor\_out(u) \to u$.

For example,

$$Factor\_out\left((x^2 + x\,y)^3\right) \to x^3(x+y)^3,$$

$$Factor\_out(a\,(b + b\,x)) \to a\,b\,(1 + x),$$

$$Factor\_out\left(2^{1/2} + 2\right) \to 2^{1/2}(1 + 2^{1/2}),$$

$$Factor\_out(a\,b\,x + a\,c\,x + b\,c\,x) \to (a\,b + a\,c + b\,c)\,x,$$

$$Factor\_out(a/x + b/x) \to a/x + b/x.$$

In the last example, $1/x$ is not isolated because the *Common_factors* operator in FO-3 retrieves only powers with positive rational exponents (see CF-4). Give a procedure for this operator.

(c) Let u be an algebraic expression in rationalized form. Give a procedure for the operator $Cancel(u)$ that is defined by the following transformation rule.

Let $n = Numerator(u)$ and $d = Denominator(u)$. Then

$$Cancel(u) \to Factor\_out(n)/Factor\_out(d).$$

For example,

$$Cancel\left(\frac{(a+b)\,c + (a+b)\,d}{a\,e + b\,e}\right) \to \frac{c+d}{e}. \qquad (6.60)$$

Note: $Cancel(u)$ does not remove all explicit common factors. For example, although

$$\frac{a\,(a+b) - a^2 - a\,b + r\,s + r\,t}{r^2}$$

has a common factor of r in the numerator and denominator and simplifies to $(s+t)/r$, this is not obtained with *Cancel*. However, if the expression is first transformed to rational-expanded form, the simplification is obtained with *Cancel*. On the other hand, if the input to *Cancel* in Expression (6.60) is transformed to rational-expanded form, the common factor is not removed. For further discussion of common factors in these cases and implicit common factors, see Cohen [24], Section 6.3.

7

Exponential and Trigonometric Transformations

This chapter is concerned with the manipulation of algebraic expressions that contain exponential or trigonometric functions. In Section 7.1 we describe expansion algorithms that expand these functions with respect to their arguments. These algorithms obtain the transformations

$$\exp(2\,x + y) \quad \to \quad (\exp(x))^2 \, \exp(y), \tag{7.1}$$

$$\sin(2\,x + y) \quad \to \quad 2\,\cos(y)\sin(x)\cos(x) + 2\,\sin(y)\,(\cos(x))^2 \tag{7.2}$$

$$- \sin(y).$$

In Section 7.2 we describe contraction algorithms that invert the transformations in (7.1) and (7.2). In addition, we describe a simplification algorithm that can verify a large class of trigonometric identities.

7.1 Exponential and Trigonometric Expansion

In this section we describe algorithms that expand the exponential and trigonometric functions that appear in an expression.

Exponential Expansion

Let u, v, and w be algebraic expressions. The exponential function satisfies the following properties[1]:

$$\exp(u + v) = \exp(u)\exp(v), \qquad (7.3)$$
$$\exp(w\,u) = \exp(u)^w. \qquad (7.4)$$

The operation that applies these transformations in a left to right manner is called *exponential expansion*, and the operation that applies the transformations in a right to left manner is called *exponential contraction*[2]. In this section we describe procedures for exponential expansion. Procedures for exponential contraction are described in Section 7.2.

The goal of exponential expansion is described in the next definition.

Definition 7.1. *An algebraic expression u is in* **exponential-expanded** *form if the argument of each exponential function in u*

1. *is not a sum;*

2. *is not a product with an operand that is an integer.*

Although Equation (7.4) provides a way to remove any operand of a product from the argument of an exponential function, it doesn't specify which operand should be removed. To eliminate this ambiguity, we only remove an integer operand from the argument[3]. This point is illustrated in the next two examples.

Example 7.2. Consider the manipulation

$$\begin{aligned}
\exp(2\,w\,x + 3\,y\,z) &= \exp(2\,w\,x)\exp(3\,y\,z) \\
&= \exp(w\,x)^2 \exp(y\,z)^3.
\end{aligned}$$

---

[1] Property (7.3) is valid in either a real number of complex number context. Property (7.4) is valid in a real context but is only valid in a complex context when w is an integer. For example, if $u = (3/2)\pi\imath$ (where $\imath = \sqrt{-1}$) and $w = 1/2$, by using the principal value of the square root function, we have $(\exp(u))^w = \sqrt{2}/2 - \sqrt{2}/2\imath$ and $\exp(w\,u) = -\sqrt{2}/2 + \sqrt{2}/2\imath$. For a discussion of the exponent relationships in a complex setting, see Pennisi [78], pages 112-113. (Implementation: Maple (mws), Mathematica (nb), MuPAD (mnb).)

[2] During automatic simplification the Mathematica system transforms the function form Exp[u] to the power E^u and also applies the contraction $E^u E^v \rightarrow E^{u+v}$. In addition, it applies the contraction $(E^u)^n \rightarrow E^{n\,u}$ when n is an integer. Therefore, to implement the expansion and contraction procedures described in this chapter in this system, it is necessary to use another representation for the exponential function. (Implementation: Mathematica (nb).)

[3] In Maple and Mathematica, an integer operand in a product is the first operand. In MuPAD, an integer operand in a product is the last operand.

The exponential on the left has an operand that is a sum and so is not in exponential-expanded form. Applying Equation (7.3), we obtain two new exponentials, which are also not in expanded form. Applying Equation (7.4) to each exponential, we obtain the expanded form of the expression. $\square$

Example 7.3. Consider the manipulation

$$
\begin{aligned}
\exp(\,2\,(x+y)) &= \exp(x+y)^2 \\
&= \exp(x)^2 \exp(y)^2.
\end{aligned}
$$

The exponential on the left has an operand that is a product with an integer operand and so is not in exponential-expanded form. Applying Equation (7.4), we obtain a new exponential that has an operand that is a sum and so is not in expanded form. Applying Equation (7.3), we obtain the expanded form of the expression. $\square$

A procedure that transforms an expression to exponential-expanded form is given in Figure 7.1. At line 4, the *Map* operator calls on the procedure recursively to search all operands of the expression for exponentials. If the resulting expression is an exponential (line 5), the procedure attempts to apply Equation (7.3) (lines 7-9) or Equation (7.4) (lines 10-13). At line 9, two new exponentials are created that may not be in expanded form, and so the procedure is called recursively to reapply the rules. (This recursion is needed in Example 7.2.) For the same reason, the procedure is applied recursively[4] to the new exponential created in line 13. (This recursion is needed in Example 7.3.)

Unfortunately, because recursion is used in two ways, to traverse all operands of the expression tree and to reapply the rules to newly created exponentials, *Expd_exp* creates some redundant recursion. To see how this happens, consider the expansion of the expression $\exp(2\,w\,x + 3\,y\,z)$ described in Example 7.2. In this case, there are 28 procedure calls with the following inputs:

$$
\begin{aligned}
\exp(2\,w\,x + 3\,y\,z),\ 2\,w\,x + 3\,yz,\ 2\,w\,x,\ 2,\ w,\ x,\ 3\,y\,z,\ 3,\ y,\ z, &\qquad (7.5) \\
\exp(2\,w\,x),\ 2\,w\,x,\ 2,\ w,\ x, &\qquad (7.6) \\
\exp(w\,x),\ w\,x,\ w,\ x, &\qquad (7.7) \\
\exp(3\,y\,z),\ 3\,y\,z,\ 3,\ y,\ z, &\qquad (7.8) \\
\exp(y\,z),\ y\,z,\ y,\ z. &\qquad (7.9)
\end{aligned}
$$

---

[4]In the Maple and MuPAD systems an expression like $\exp(2\,(x+y))$ is transformed to $\exp(2\,x + 2\,y)$ by automatic simplification, and so the recursive call at line 13 is not needed. (Implementation: Maple (mws), MuPAD (mnb).)

Procedure *Expd_exp(u)*;
Input
    u : an algebraic expression;
Output
    an algebraic expression in exponential-expanded form;
Local Variables
    v, A, f;
Begin
1   **if** *Kind(u)* ∈ {**integer, fraction, symbol**} **then**
2   *Return(u)*
3   **else**
4   *v := Map(Expd_exp, u)*;
5   **if** *Kind(v)* = exp **then**
6   *A := Operand(v,1)*;
7   **if** *Kind(A)* = " + " **then**
8   *f := Operand(A, 1)*;
9   *Return(Expd_exp(exp(f)) * Expd_exp(exp(A − f)))*
10   **elseif** *Kind(A)* = " * " **then**
11   *f := Operand(A, 1)*;
12   **if** *Kind(f)* = **integer then**
13   *Return(Expd_exp(exp(A/f))^f)*;
14   *Return(v)*
End

Figure 7.1. An MPL procedure that transforms an algebraic expression to exponential-expanded form. (Implementation: Maple (txt), Mathematica (txt), MuPAD (txt). In the MuPAD implementation, the statement at line 11 assigns the last operand of A to f.)

The inputs associated with the tree traversal of $\exp(2\,w\,x + 3\,y\,z)$ from line 4 are given in (7.5). Since there are no exponentials in $2\,w\,x + 3\,y\,z$, the next recursive step occurs when the new exponential $\exp(2\,w\,x)$ is created at line 9, which leads to the inputs for the next sequence of calls in (7.6). Observe that redundant recursion occurs (from line 4) because all subexpressions of $2\,w\,x$ are traversed for a second time. In a similar way, the next sequence of inputs is given in (7.7) when the procedure attempts at line 13 to expand the new exponential $\exp(w\,x)$. Once again, more redundant recursion arises (from line 4) as the sub-expressions of $w\,x$ are traversed for a third time. Finally, more redundant recursion occurs (from lines 9 and 13) with expansion of the new expressions $\exp(3\,y\,z)$ and $\exp(y\,z)$ (see (7.8) and (7.9)).

One simple way to eliminate the redundant recursion is to implement the procedure in a language that remembers the input-output values of procedure calls. Another approach is to separate the two roles for recursion by using two procedures. The procedures that perform exponential expansion in this way are shown in Figure 7.2. Notice that there is an outer main procedure *Expand_exp* and an inner procedure *Expand_exp_rules*. The recursion that is used to traverse all operands of the expression tree is obtained with the *Map* operator at line 4 of *Expand_exp*. This procedure also calls on *Expand_exp_rules* at line 6, which takes as input the argument of an exponential function and applies the transformation rules (Equations (7.3) and (7.4)). Notice that the reapplication of the rules is obtained in *Expand_exp_rules* at lines 3 and 7. Since *Expand_exp_rules* only applies recursion when a rule is applied, some redundant recursion is eliminated. For example, to obtain the expanded form of $\exp(2\,w\,x + 3\,y\,z)$, the sequence of inputs to *Expand_exp* is still given in (7.5), while the sequence of inputs for *Expand_exp_rules* is given by

$$2\,w\,x + 3\,y\,z, \quad 2\,w\,x, \quad w\,x, \quad 3\,y\,z, \quad y\,z.$$

Using the two procedures in Figure 7.2, there are 15 procedure calls, while using the single procedure in Figure 7.1, there are 28 procedure calls.

Appraisal of *Expand_exp*. In the present form, the algorithm encounters a division by zero whenever an application of a transformation rule together with automatic simplification transforms a denominator to zero. This occurs, for example, with

$$\frac{1}{\exp(2\,x) - \exp(x)^2}.$$

A modification of the algorithm that recognizes this and returns the symbol **Undefined** is described in Exercise 2.

Since *Expand_exp* is applied in the simplification context of automatic simplification, it is unable to obtain some transformations that require additional algebraic operations. For example, the manipulation

$$\exp((x + y)\,(x - y)) = \exp(x^2)/\exp(y^2) \tag{7.10}$$

is not obtained with exponential expansion unless the argument of the exponential on the left is first algebraically expanded. A modification of the algorithm that obtains this transformation is described in Exercise 3.

Procedure *Expand_exp(u)*;
Input
    u : an algebraic expression;
Output
    an algebraic expression in exponential-expanded form;
Local Variables
    v;
Begin
1   **if** *Kind(u)* $\in$ {**integer, fraction, symbol**} **then**
2   *Return(u)*
3   **else**
4   *v := Map(Expand_exp, u)*;
5   **if** *Kind(v)* = exp **then**
6   *Return(Expand_exp_rules(Operand(v, 1)))*
7   **else**
8   *Return(v)*
End

Procedure *Expand_exp_rules(A)*;
Input
    A : an algebraic expression that is the argument of an exponential
        function;
Output
    the exponential-expanded form of exp(*A*);
Local Variables
    f;
Begin
1   **if** *Kind(A)* = " + " **then**
2   *f := Operand(A, 1)*;
3   *Return(Expand_exp_rules(f) * Expand_exp_rules(A − f))*
4   **elseif** *Kind(A)* = " * " **then**
5   *f := Operand(A, 1)*;
6   **if** *Kind(f)* = **integer then**
7   *Return(Expand_exp_rules(A/f)^f)*;
8   *Return(exp(A))*
End

Figure 7.2. Two MPL procedures that separate the two roles for recursion in exponential expansion. (Implementation: Maple (txt), Mathematica (txt), MuPAD (txt). In the MuPAD implementation, the statement at line 5 of *Expand_exp_rules* assigns the last operand of *A* to *f*.)

Trigonometric Expansion

The sin and cos functions satisfy the identities:

$$\sin(\theta + \phi) = \sin(\theta)\,\cos(\phi) + \cos(\theta)\,\sin(\phi), \tag{7.11}$$
$$\cos(\theta + \phi) = \cos(\theta)\,\cos(\phi) - \sin(\theta)\,\sin(\phi). \tag{7.12}$$

The *trigonometric expansion* operation applies these identities in a left to right manner to all sin and cos functions in an expression. We also obtain expanded forms for $\sin(n\,\theta)$ and $\cos(n\,\theta)$ (n a positive integer) by viewing the argument $n\,\theta$ as a sum with n identical operands θ and repeatedly applying the rules. In addition, by applying the identities $\sin(-\theta) = -\sin(\theta)$ and $\cos(-\theta) = \cos(\theta)$, we obtain expanded forms for $\sin(n\theta)$ and $\cos(n\theta)$ when n is a negative integer as well.

The goal of trigonometric expansion is described in the next definition.

Definition 7.4. *An expression u is in* **trigonometric-expanded** *form if the argument of each* sin *and* cos *function in u*

1. *is not a sum;*

2. *is not a product with an operand that is an integer.*

The definition is given only in terms of sine and cosine functions because the other trigonometric functions can be expressed in terms of these functions. (See the *Trig_substitute* procedure in Figure 5.12 on page 190.)

Example 7.5. Consider the manipulation

$$\begin{aligned}
\sin(2\,x + 3\,y) &= \sin(2\,x)\cos(3\,y) + \cos(2\,x)\sin(3\,y) \tag{7.13}\\
&= 2\sin(x)\cos(x)\left(\cos^3(y) - 3\cos(y)\sin^2(y)\right) \tag{7.14}\\
&\quad + \left(\cos^2(x) - \sin^2(x)\right)\left(3\cos^2(y)\sin(y) - \sin^3(y)\right).
\end{aligned}$$

The sin on the left is not in trigonometric-expanded form because its argument is a sum. Applying the identity (7.11), we obtain two new sines and two new cosines that are also not in expanded form. By reapplying the rules, we obtain the final expanded form in Expression (7.14). □

Example 7.6. Consider the manipulation

$$\begin{aligned}
\sin(2\,(x+y)) &= 2\,\sin(x+y)\cos(x+y) \tag{7.15}\\
&= 2\,(\sin(x)\cos(y) + \cos(x)\sin(y))(\cos(x)\cos(y)\\
&\quad - \sin(x)\sin(y)). \tag{7.16}
\end{aligned}$$

The sin on the left is not in trigonometric-expanded form because its argument is a product with an integer operand. Applying the identity (7.11), we obtain a new sine and a new cosine that are not in expanded form. By reapplying the rules, we obtain the expanded form in Expression (7.16).

$\square$

Because of the identity

$$\sin^2(\theta) + \cos^2(\theta) = 1, \qquad (7.17)$$

an expression can have a number of trigonometric-expanded forms. For example, our algorithm (shown in Figure 7.3) obtains the expanded form

$$\cos(5\,x) = \cos^5(x) - 10\,\cos^3(x)\sin^2(x) + 5\,\cos(x)\sin^4(x). \qquad (7.18)$$

By using the identity (7.17), however, we can remove $\sin^2(x)$ and $\sin^4(x)$ from the expression and obtain another expanded form that involves only cosines

$$\cos(5\,x) = 16\,\cos^5(x) - 20\,\cos^3(x) + 5\,\cos(x). \qquad (7.19)$$

Although a simple expansion algorithm is obtained by repeatedly applying the identities (7.11) and (7.12), a straightforward implementation can involve excessive recursion. We describe next three modifications to this process that reduce some of this recursion.

First, as with exponential expansion, recursion is used two ways: to examine all the operands of an expression tree and to reapply the transformations (7.11) and (7.12) when a new sine or cosine is created. To reduce redundant recursion, we divide the computation into two procedures that handle each of the recursive tasks.

The next example shows another way that redundant recursion can arise.

Example 7.7. Consider the trigonometric expansion of $\sin(a + b + c + d)$. First, we apply the identity (7.11) with $\theta = a$ and $\phi = b + c + d$ to obtain

$$\sin(a + b + c + d) = \sin(a)\cos(b + c + d) + \cos(a)\sin(b + c + d).$$

Next, apply the identities (7.11) and (7.12) recursively to $\cos(b+c+d)$ and $\sin(b + c + d)$ with $\theta = b$ and $\phi = c + d$. to obtain

$$\cos(b + c + d) = \cos(b)\cos(c + d) - \sin(b)\sin(c + d),$$

$$\sin(b + c + d) = \sin(b)\cos(c + d) + \cos(b)\sin(c + d).$$

Because both of the expressions on the right require expansions for $\cos(c+d)$ and $\sin(c + d)$, the next recursive application of the rules leads to some redundant recursion. Using this approach, this example requires seven

Procedure *Expand_trig(u)*;
Input
    u: an algebraic expression;
Output
    an algebraic expression in trigonometric-expanded form;
Local Variables *v*;
Begin
1   **if** *Kind(u)* ∈ {**integer, fraction, symbol**} **then**
2   *Return(u)*
3   **else**
4        $v := Map(Expand\_trig, u)$;
5   **if** *Kind(v)* = sin **then**
6   *Return(Operand(Expand_trig_rules(Operand(v, 1)), 1))*
7   **elseif** *Kind(v)* = cos **then**
8   *Return(Operand(Expand_trig_rules(Operand(v, 1)), 2))*
9   **else**
10   *Return(v)*
End

Procedure *Expand_trig_rules(A)*;
Input
    A: an algebraic expression that is the argument of a sin or cos;
Output
    a two element list $[s, c]$ where *s* and *c* are the
    trigonometric-expanded forms of sin(*A*) and cos(*A*);
Local Variables *f, r, s, c*;
Begin
1   **if** *Kind(A)* = " + " **then**
2        $f := Expand\_trig\_rules(Operand(A, 1))$;
3        $r := Expand\_trig\_rules(A - Operand(A, 1))$;
4        $s := Operand(f, 1) * Operand(r, 2) + Operand(f, 2) * Operand(r, 1)$;
5        $c := Operand(f, 2) * Operand(r, 2) - Operand(f, 1) * Operand(r, 1)$;
6   *Return([s, c])*
7   **elseif** *Kind(A)* = " * " **then**
8        $f := Operand(A, 1)$;
9   **if** *Kind(f)* = **integer then**
10   *Return([Multiple_angle_sin(f, A/f), Multiple_angle_cos(f, A/f)])*;
11   *Return([sin(A), cos(A)])*
End

Figure 7.3. MPL procedures that transform an algebraic expression to trigonometric-expanded form. (Implementation: Maple (txt), Mathematica (txt), MuPAD (txt). In the MuPAD implementation, the statement at line 8 of *Expand_trig_rules* assigns the last operand of *A* to *f*.)

applications of the rules. In general, if a sin or cos has an argument that is a sum of n symbols, the number of rule applications grows exponentially as $2^{n-1} - 1$ (Exercise 6(a)). $\square$

There are a number of ways to eliminate this redundant recursion. One way is simply to implement the algorithm in a language that remembers the input-output values of procedure calls. Another approach, which we use here, is to obtain $\sin(A)$ and $\cos(A)$ simultaneously. (See the discussion on page 285 and the procedure *Expand_trig_rules* in Figure 7.3.) This approach requires only $2(n-1)$ rule applications to expand a sin or cos of a sum of n symbols (Exercise 6(b)).

Another improvement to the algorithm is based on the following representations for multiple angle expansions. For n a positive integer,

$$\cos(n\,\theta) = \sum_{\substack{j=0 \\ j \text{ even}}}^{n} (-1)^{j/2} \binom{n}{j} \cos^{n-j}(\theta) \sin^{j}(\theta), \qquad (7.20)$$

$$\sin(n\,\theta) = \sum_{\substack{j=1 \\ j \text{ odd}}}^{n} (-1)^{(j-1)/2} \binom{n}{j} \cos^{n-j}(\theta) \sin^{j}(\theta). \qquad (7.21)$$

For example, the expansion in Equation (7.18) is obtained using the first formula.

These representations are derived using the exponential representations for sin and cos and the binomial theorem. For example to obtain the sum (7.20),

$$
\begin{aligned}
\cos(n\,\theta) &= \frac{\exp(\imath\,n\,\theta) + \exp(-\imath\,n\,\theta)}{2} = \frac{\exp(\imath\,\theta)^{n} + \exp(-\imath\,\theta)^{n}}{2} \\
&= \frac{(\cos(\theta) + \imath \sin(\theta))^{n} + (\cos(\theta) - \imath \sin(\theta))^{n}}{2} \\
&= (1/2)\left(\sum_{j=0}^{n} \binom{n}{j} \cos^{n-j}(\theta) \imath^{j} \sin^{j}(\theta) \right. \\
&\qquad\qquad \left. + \sum_{j=0}^{n} \binom{n}{j} \cos^{n-j}(\theta)(-\imath)^{j} \sin^{j}(\theta) \right) \\
&= (1/2)\sum_{j=0}^{n} \binom{n}{j} \cos^{n-j}(\theta) \sin^{j}(\theta)\, \imath^{j} (1 + (-1)^{j}).
\end{aligned}
$$

However, using

$$i^j(1 + (-1)^j) = \begin{cases} 2\,(-1)^{j/2}, & j \text{ even,} \\ 0, & j \text{ odd,} \end{cases}$$

we obtain the representation (7.20).

Another approach for expanding $\sin(n\,\theta)$ and $\cos(n\,\theta)$ that uses recurrence relations is described in Exercise 10.

Procedures that transform an expression to trigonometric-expanded form are given in Figure 7.3. The main procedure *Expand_trig* applies the process to the operands of an expression using the *Map* operator in line 4 and, if the resulting expression is a sin or cos, invokes the procedure *Expand_trig_rules* to apply the expansion rules (lines 6 and 8).

Expand_trig_rules(A) returns a two element list with the trigonometric-expanded forms of $\sin(A)$ and $\cos(A)$. When A is a sum, the procedure is applied recursively to both *Operand*($A, 1$) and to $A - $ *Operand*($A, 1$) (line 2-3) after which the identities (7.11) and (7.12) are applied to the resulting expressions (lines 4-5). (This recursion is needed in Example 7.5.) When A is a product with an integer operand, the procedure invokes the *Multiple_angle_sin* and *Multiple_angle_cos* procedures, which apply the multiple angle representations given in (7.21) and (7.20). (These procedures are left to the reader (Exercise 5).) This step is also recursive because these procedures invoke *Expand_trig_rules*. (This recursion is needed in Example 7.7.) Finally, when neither transformation rule applies, the procedure returns $[\sin(A), \cos(A)]$ (line 11).

Appraisal of *Expand_trig*. In the present form the algorithm encounters a division by zero whenever an application of a transformation rule together with automatic simplification transforms a denominator to zero. For example, this occurs with the expression $1/(\sin(2\,x) - 2\sin(x)\cos(x))$. A modification of the algorithm that recognizes this and returns **Undefined** is described in Exercise 7.

Since the *Expand_trig* algorithm does not include algebraic expansion, it misses some opportunities to apply the trigonometric expansion rules. For example, the expression $\sin((x + y)^2)$ is not expanded because $(x + y)^2$ is not in (algebraic) expanded form. In addition, the output for an expression like $\sin(a + b + c + d)$ is cumbersome because it is returned in a nested form rather than in an algebraic expanded form. In Exercise 8, we describe modifications of the procedures that handle these problems.

In some instances, *Expand_trig* distorts the mathematical meaning of an expression. Consider the expression

$$\frac{\sin(2\,x) - 2\,\sin(x)\cos(x)}{(\sin(x))^2 + (\cos(x))^2 - 1}.$$

Strictly speaking, this expression is an indeterminate form because both the numerator and denominator simplify to 0. However, *Expand_trig* simplifies the expression to 0 because the expanded form of the numerator is 0 while the denominator is already in expanded form and is not changed. In Section 7.2, we describe the *Simplify_trig* operator that recognizes the problem for this expression and indicates that it is undefined. However, because it is theoretically impossible to give an algorithm that can always determine if an algebraic expression simplifies to 0, it is impossible to avoid this problem in all cases[5].

Exercises

1. Give the exponential-expanded form of the expression

$$\exp((\exp(2\,x) - \exp^2(x) + 1)\,(2\,x + 3\,y)).$$

2. The *Expand_exp* algorithm encounters a division by zero if the transformation rules transform a sub-expression in a denominator to zero. For example, this occurs with $1/(\exp(2\,x) - \exp(x)^2)$. Modify the procedure so that it recognizes this situation and returns the global symbol **Undefined** when it occurs.

3. Modify the definition of an exponential-expanded expression u given in Definition 7.1 so that it includes properties 1 and 2 from that definition as well as the property that each complete sub-expression of u is in algebraic expanded form. Modify the procedures in Figure 7.2 to obtain an expression in this form. For example, your procedures should obtain

$$Expand\_exp\left(\exp\left((x + y)^2\right)\right) \rightarrow \exp\left(x^2\right)(\exp(x\,y))^2\exp\left(y^2\right).$$

4. Let u, v, and w be algebraic expressions. The natural logarithm function satisfies the following two properties:

$$\ln(u\,v) = \ln(u) + \ln(v), \qquad\qquad (7.22)$$
$$\ln(u^w) = w\,\ln(u). \qquad\qquad (7.23)$$

An algebraic expression is in *log-expanded* form if the argument of each logarithm is not a product or a power. For example, the manipulation

$$\ln((w\,x)^a) + \ln(y^b\,z) \rightarrow a\left(\ln(w) + \ln(x)\right) + b\ln(y) + \ln(z).$$

transforms the expression on the left to log-expanded form. An expression can be transformed to log-expanded form by applying Equations (7.22) and (7.23) in a left to right manner. Give a procedure *Expand_log(u)* that transforms an algebraic expression u to log-expanded form.

---

[5] See footnote 6 on page 145.

5. Let θ be an algebraic expression, and let n be an integer. Give procedures

$$Multiple\_angle\_sin(n, \theta), \qquad Multiple\_angle\_cos(n, \theta)$$

that find the expansions for $\sin(n\,\theta)$ and $\cos(n\,\theta)$ using Equations (7.20) and (7.21). Keep in mind when θ is a sum, $\sin(\theta)$ and $\cos(\theta)$ are not in expanded form. (This situation occurs in Example 7.6.) In this case it is necessary[6] to expand these expressions with *Expand_trig_rules*.

Be sure to account for the possibility that n is a negative integer. Although most computer algebra systems apply transformations such as $\sin(-2\,y) \rightarrow -\sin(2\,y)$ during automatic simplification, this situation will arise in our algorithm with $\sin(x - 2\,y)$ because *Expand_trig_rules* takes the *argument* of a sin or cos as input rather than the function form.

6. Consider the expansion of $\sin(A)$ and $\cos(A)$ where A is a sum of n symbols.

   (a) Show that to expand $\sin(A)$ or $\cos(A)$ using the approach in Example 7.7 requires $2^{n-1} - 1$ rule applications.

   (b) Show that to expand $\sin(A)$ and $\cos(A)$ simultaneously using the algorithm in Figure 7.3 requires $2\,(n - 1)$ rule applications.

7. The *Expand_trig* algorithm may encounter a division by zero if the transformation rules transform a sub-expression in a denominator to zero. For example, this occurs with $1/(\sin(2\,x) - 2\sin(x)\,\cos(x))$. Modify the procedure so that it recognizes this situation and returns the global symbol **Undefined** when it occurs.

8. Suppose we modify the definition of a trigonometric-expanded expression u in Definition 7.4 so that it includes properties (1) and (2) in that definition as well as the property that each complete sub-expression of u is in algebraic expanded form. Modify the procedures in Figure 7.3 to obtain an expression in this form. For example, your procedures should obtain

$$Expand\_trig\left(\sin\left((x + y)^2\right)\right) \rightarrow$$
$$\sin(x^2)\left(\left((\cos(x\,y))^2 - (\sin(x\,y))^2\right)\cos(y^2)\right.$$
$$\left. - 2\,\cos(x\,y)\sin(x\,y)\sin(y^2)\right)$$
$$+ \cos(x^2)\left(2\,\cos(x\,y)\sin(x\,y)\cos(y^2)\right.$$
$$\left. + \left((\cos(x\,y))^2 - (\sin(x\,y))^2\right)\sin(y^2)\right).$$

9. In this exercise we describe an extension to the *Expand_trig* algorithm to include the sinh and cosh functions. These functions satisfy the identities

$$\sinh(\theta + \phi) = \sinh(\theta)\,\cosh(\phi) + \cosh(\theta)\,\sinh(\phi),$$
$$\cosh(\theta + \phi) = \cosh(\theta)\,\cosh(\phi) + \sinh(\theta)\,\sinh(\phi),$$

---

[6]In the Maple and MuPAD systems, it is not necessary to invoke *Expand_trig_rules* here because an integer is distributed over the operands of a sum by automatic simplification, and so θ cannot be a sum.

$$\sinh(-\theta) = -\sinh(\theta), \quad \cosh(-\theta) = \cosh(\theta),$$
$$\cosh(n\,\theta) \pm \sinh(n\,\theta) = (\cosh(\theta) \pm \sinh(\theta))^n, \tag{7.24}$$

where n is a positive integer.

(a) Using the identity (7.24) it follows that

$$\cosh(n\,\theta) = 1/2\left((\cosh(\theta) + \sinh(\theta))^n + (\cosh(\theta) - \sinh(\theta))^n\right),$$

$$\sinh(n\,\theta) = 1/2\left((\cosh(\theta) + \sinh(\theta))^n - (\cosh(\theta) - \sinh(\theta))^n\right).$$

Use these formulas to derive representations similar to those in Equations (7.20) and (7.21) for $\sinh(n\,\theta)$ and $\cosh(n\,\theta)$.

(b) Extend the *Expand_trig* algorithm so that it also expands the sinh and cosh functions.

10. This exercise describes another approach that finds the expanded form for $\sin(n\,\theta)$ and $\cos(n\,\theta)$ that uses recurrence relations.

(a) Show that for $n \geq 2$, $p_n = \sin(n\,\theta)$ satisfies the recurrence relation

$$p_n = 2\cos(\theta)\,p_{n-1} - p_{n-2}, \quad p_1 = \sin(\theta), \quad p_0 = 0.$$

(b) Show that for $n \geq 2$, $q_n = \cos(n\,\theta)$ satisfies the recurrence relation

$$q_n = 2\cos(\theta)\,q_{n-1} - q_{n-2}, \quad q_1 = \cos(\theta), \quad q_0 = 1.$$

Notice that this recurrence relation gives the expansion in (7.19).

(c) Give procedures that find the expanded forms for $\sin(n\,\theta)$ and $\cos(n\,\theta)$ using the recurrence relations in parts (a) and (b). Be sure to account for the possibility that n is negative and θ is a sum. (See the discussion in Exercise 5 above.)

11. In this exercise we ask you to give a procedure for trigonometric expansion of the tangent function that is based on the identity

$$\tan(\theta + \phi) = \frac{\tan(\theta)\,\tan(\phi)}{1 - \tan(\theta)\,\tan(\phi)}. \tag{7.25}$$

(a) Let n be a positive integer. Show that

$$\tan(n\,\theta) = \frac{\displaystyle\sum_{\substack{j=1 \\ j\ \text{odd}}}^{n} (-1)^{(j-1)/2}\binom{n}{j}\tan^j(\theta)}{\displaystyle\sum_{\substack{j=0 \\ j\ \text{even}}}^{n} (-1)^{j/2}\binom{n}{j}\tan^j(\theta)}. \tag{7.26}$$

(b) Give a procedure *Expand_tan(u)* that is based on Equations (7.25) and (7.26).

7.2 Exponential and Trigonometric Contraction

In this section we describe the exponential and trigonometric contraction operators and a trigonometric simplification operator that can verify a large class of trigonometric identities.

Exponential Contraction

Exponential contraction applies the two transformation rules[7]

$$\exp(u)\,\exp(v) \rightarrow \exp(u+v), \tag{7.27}$$

$$\exp(u)^w \rightarrow \exp(w\,u). \tag{7.28}$$

The goal of this operation is described in the following definition.

Definition 7.8. *An algebraic expression u is in* **exponential-contracted** *form if it satisfies the following properties.*

1. *Each product in u contains at most one operand that is an exponential function.*

2. *Each power in u does not have an exponential function for its base.*

3. *Each complete sub-expression of u is in algebraic-expanded form.*

Properties (1) and (2) are obtained by applying the transformations (7.27) and (7.28). We have included property (3) because algebraic expansion creates new opportunities to apply these rules. This point is illustrated in the next example.

Example 7.9. Consider the manipulation

$$
\begin{aligned}
\exp(x)\,(\exp(x) + \exp(y)) &= (\exp(x))^2 + \exp(x)\,\exp(y) \\
&= \exp(2\,x) + \exp(x+y).
\end{aligned}
$$

The expression on the left is not in contracted form because it is not in algebraic-expanded form. Algebraic expansion gives a new sum with two operands, a new power and a new product, that are not in contracted form. Applying the transformations (7.27) and (7.28) we obtain the contracted form. □

---

[7]See footnote 1 on page 276 for some remarks about the validity of these transformations in real and complex contexts.

Procedure   *Contract_exp(u)*;
Input
    u : an algebraic expression;
Output
    an algebraic expression in exponential-contracted form;
Local Variables
    v;
Begin
1   **if** *Kind(u)* ∈ {**integer, fraction, symbol**} **then**
2   *Return(u)*
3   **else**
4   *v := Map(Contract_exp, u)*;
5   **if** *Kind(v)* ∈ {" ∗ "," ∧ "} **then**
6   *Return(Contract_exp_rules(v))*
7   **else**
8   *Return(v)*
End

Figure 7.4. The main MPL procedure that transforms an algebraic expression to exponential-contracted form. (Implementation: Maple (txt), Mathematica (txt), MuPAD (txt).)

Example 7.10. Consider the manipulation

$$\exp(\exp(x))^{\exp(y)} \;=\; \exp(\exp(x)\,\exp(y))$$
$$=\; \exp(\exp(x+y)).$$

The expression on the left is not in contracted form because it is a power with an exponential for a base. Applying Equation (7.28) we obtain an expression with a new product that is not in contracted form. Applying Equation (7.27) we obtain the contracted form.   □

Procedures[8] for exponential contraction are shown in Figures 7.4 and 7.5. Notice that there is an outer main procedure *Contract_exp* and an inner procedure *Contract_exp_rules*. We have divided the computation in this way to account for the two types of recursion that occur in the algorithm and to indicate clearly where algebraic expansion or a reapplication of the rules is required.

The recursion that is used to traverse all operands of the expression tree is obtained with the *Map* operator in line 4 of *Contract_exp*. At line 6, this

---

[8]See footnote 2 on page 276 concerning the Mathematica implementation of these procedures.

Procedure *Contract_exp_rules*(u);
Input
    u : an algebraic expression that is sent by either *Contract_exp*
        or a recursive call of this procedure;
Output
    an algebraic expression in exponential-contracted form;
Local Variables v, b, s, p, i, y;
Begin

```
1     v := Expand_main_op(u);
2     if  Kind(v) = " ∧ " then
3         b := Operand(v, 1);
4         s := Operand(v, 2);
5         if  Kind(b) = exp then
6             p := Operand(b, 1) * s;
7             if  Kind(p) ∈ {" * "," ∧ "} then
8                 p := Contract_exp_rules(p);
9             Return(exp(p))
10        else
11            Return(v)
12    elseif  Kind(v) = " * " then
13        p := 1;
14        s := 0;
15        for  i := 1 to  Number_of_operands(v) do
16            y := Operand(v, i);
17            if  Kind(y) = exp then
18                s := s + Operand(y, 1)
19            else
20                p := p * y;
21        Return(exp(s) * p)
22    elseif  Kind(v) = " + " then
23        s := 0;
24        for  i := 1 to  Number_of_operands(v) do
25            y := Operand(v, i);
26            if  Kind(y) ∈ {" * "," ∧ "} then
27                s := s + Contract_exp_rules(y)
28            else
29                s := s + y;
30        Return(s)
31    else
32        Return(v)
End
```

Figure 7.5. The inner MPL procedure for exponential contraction. (Implementation: Maple (txt), Mathematica (txt), MuPAD (txt).)

procedure calls on *Contract_exp_rules* which applies algebraic expansion and the transformation rules (Equations (7.27) and (7.28)). Notice that we only invoke *Contract_exp_rules* when v is a product or a power.

The second type of recursion occurs when either algebraic expansion or an application of one of the contraction rules creates a new sum, product, or power that is not in contracted form. This recursion is invoked at lines 8 and 27 of *Contract_exp_rules*. At line 1 we algebraically expand the input expression. To avoid redundant recursion, we use the *Expand_main_op* operator that does not recursively expand its operands (Exercise 6, page 258). When v is power (line 2) with an exponential function at its base (line 5), we apply Equation (7.28) to obtain a new operand p of the exponential. Then, if p is a product or a power (by automatic simplification), we recursively contract this expression (lines 7 and 8). (This recursion is required in Example 7.10.) If the base is not an exponential, no transformation is possible and we return v (line 11).

Next, if v is a product (line 12), we loop through its operands (lines 13-20) combining exponentials with Equation (7.27). Lines 22-30 handle the case when v is a sum. (This part is only invoked when a sum is created by the expansion in line 1.) In this case, we loop through the operands and recursively contract when an operand is a product or power. Finally, lines 31 and 32 apply to any other type of expression that was created by the expansion at line 1. For example, for $u = (\sqrt{2}+1)(\sqrt{2}-1)$ the expansion at line 1 assigns 1 to v which is returned at line 32.

Appraisal of *Contract_exp*. Although exponential contraction acts as an expression simplifier for many expressions with exponentials, it does not simplify all such expressions. For example, consider the exponential contraction

$$Contract\_exp\left(\frac{1}{\exp(x)\,(\exp(y)+\exp(-x))} - \frac{\exp(x+y)-1}{(\exp(x+y))^2 - 1}\right) \quad (7.29)$$

$$\rightarrow \frac{\exp(-x)}{\exp(y)+\exp(-x)} - \frac{\exp(x+y)}{(\exp(x+y))^2 - 1} + \frac{1}{(\exp(x+y))^2 - 1}.$$

Although the (uncontracted) expression in (7.29) simplifies to 0, this simplification is not obtained with *Contract_exp*. There are two reasons for this. First, since the first term in (7.29) has the internal form $\exp(x)^{-1}\,(\exp(y)+\exp(-x))^{-1}$ which is in algebraic-expanded form, the contraction operation does not distribute $\exp(x)$ over the sum $\exp(y) + \exp(-x)$. Next, in the second term in (7.29), the numerator and denominator have a common factor $\exp(x+y) - 1$ that is not eliminated by exponential contraction. One way to simplify the expression in (7.29) is to first rationalize it using the

Rationalize_expression operator described in Section 6.5 and then contract the numerator of the resulting expression. An operator that obtains the simplification in this way is described in Exercise 4.

Trigonometric Contraction

The sin and cos functions satisfy the identities:

$$\sin(\theta)\sin(\phi) = \frac{\cos(\theta - \phi)}{2} - \frac{\cos(\theta + \phi)}{2}, \tag{7.30}$$

$$\cos(\theta)\cos(\phi) = \frac{\cos(\theta + \phi)}{2} + \frac{\cos(\theta - \phi)}{2}, \tag{7.31}$$

$$\sin(\theta)\cos(\phi) = \frac{\sin(\theta + \phi)}{2} + \frac{\sin(\theta - \phi)}{2}. \tag{7.32}$$

The *trigonometric contraction* operation applies these identities in a left to right manner. By repeatedly applying the identities (7.30) and (7.31), we also obtain contracted forms for $\sin^n(\theta)$ and $\cos^n(\theta)$ (for an integer $n > 1$). The goal of this operation is given in the following definition.

Definition 7.11. *An expression u is in* **trigonometric-contracted** *form if it satisfies the following properties.*

1. *A product in u has at most one operand that is a sine or cosine.*

2. *A power in u with a positive integer exponent does not have a base that is a sine or cosine.*

3. *Each complete sub-expression of u is in algebraic-expanded form.*

Notice that the definition does not refer to the tan, cot, sec, and csc functions because these functions can be expressed in terms of sin and cos. We have included property (3) because algebraic expansion creates new opportunities to apply the contraction rules. This point is illustrated in the next two examples.

Example 7.12. Consider the manipulation

$$(\sin(x) + \cos(y))\cos(y) = \sin(x)\cos(y) + \cos^2(y)$$

$$= \frac{\sin(x + y)}{2} + \frac{\sin(x - y)}{2} + \frac{1}{2} + \frac{\cos(2y)}{2}.$$

The expression on the left is not in contracted form because it is not in algebraic-expanded form. By expanding the expression, we obtain a new sum with two operands, a new product and a new power, that are not in contracted form. By applying the identities (7.32) and (7.31), we obtain the contracted form. □

Example 7.13. Consider the manipulation

$$\sin^2(x)\cos^2(x) = \left(\frac{1}{2} - \frac{\cos(2x)}{2}\right)\left(\frac{1}{2} + \frac{\cos(2x)}{2}\right) \tag{7.33}$$

$$= \frac{1}{4} - \frac{\cos^2(2x)}{4} \tag{7.34}$$

$$= \frac{1}{4} - \frac{1/2 + \cos(4x)/2}{4}$$

$$= \frac{1}{8} - \frac{\cos(4x)}{8}.$$

The expression on the left is not in contracted form because it contains a sine and cosine to positive integer powers. Applying the identities (7.30) and (7.31), we obtain a new product that is not in contracted form because it is not in algebraic-expanded form. Algebraically expanding the right side of Equation (7.33), we obtain in (7.34) a new sum that again is not in contracted form because it contains a positive integer power of a cosine. Applying the identity (7.31) and algebraically expanding, we obtain the contracted form. □

A simple algorithm for trigonometric contraction is obtained by repeatedly applying the identities (7.30), (7.31), and (7.32) although the approach involves an excessive amount of recursion. As with exponential contraction, we can reduce the redundant recursion by dividing the operation into two procedures. Another improvement involves the contraction of positive integer powers of sines and cosines using the following representations. For n a positive integer,

$$\cos^n(\theta) = \begin{cases} \dfrac{\binom{n}{n/2}}{2^n} + \dfrac{1}{2^{n-1}}\displaystyle\sum_{j=0}^{n/2-1}\binom{n}{j}\cos((n-2j)\theta), & n \text{ even}, \\ \dfrac{1}{2^{n-1}}\displaystyle\sum_{j=0}^{\lfloor n/2\rfloor}\binom{n}{j}\cos((n-2j)\theta), & n \text{ odd}, \end{cases} \tag{7.35}$$

$$\sin^n(\theta) =$$

$$
\begin{cases}
\dfrac{(-1)^n \binom{n}{n/2}}{2^n} + \dfrac{(-1)^{\frac{n}{2}}}{2^{n-1}} \displaystyle\sum_{j=0}^{n/2-1} (-1)^j \binom{n}{j} \cos((n-2j)\theta), & n \text{ even}, \\[3em]
\dfrac{(-1)^{\frac{n-1}{2}}}{2^{n-1}} \displaystyle\sum_{j=0}^{\lfloor n/2 \rfloor} \binom{n}{j}(-1)^j \sin((n-2j)\theta), & n \text{ odd},
\end{cases}
$$

$$(7.36)$$

where the floor function $\lfloor n/2 \rfloor$ is the largest integer $\leq n/2$. We verify the representation for $\cos^n(\theta)$ for n odd. Using the exponential representation for $\cos(\theta)$ and the binomial theorem, we have

$$
\begin{aligned}
\cos^n(\theta) &= \left(\frac{e^{i\theta} + e^{-i\theta}}{2}\right)^n \\[1em]
&= \frac{1}{2^n} \sum_{j=0}^{n} \binom{n}{j}(e^{i\theta})^{n-j}(e^{-i\theta})^j \\[1em]
&= \frac{1}{2^n} \sum_{j=0}^{n} \binom{n}{j} e^{i(n-2j)\theta} \\[1em]
&= \frac{1}{2^n}\left(\sum_{j=0}^{\lfloor n/2 \rfloor} \binom{n}{j} e^{i(n-2j)\theta} + \sum_{j=\lfloor n/2 \rfloor+1}^{n} \binom{n}{j} e^{i(n-2j)\theta} \right),
\end{aligned}
$$

where the two sums in the last expression have the same number of terms. We can combine these two sums by expressing the second sum in terms of a new summation index $k = n - j$. Observe that since n is odd, $n = 2\lfloor n/2 \rfloor + 1$, and this implies that $k = \lfloor n/2 \rfloor$ when $j = \lfloor n/2 \rfloor + 1$. Therefore, by reversing the order of summation in the second sum and using the identity

$$\binom{n}{n-k} = \binom{n}{k},$$

we have

$$
\begin{aligned}
\cos^n(\theta) &= \frac{1}{2^n}\left(\sum_{j=0}^{\lfloor n/2 \rfloor} \binom{n}{j} e^{i(n-2j)\theta} + \sum_{k=0}^{\lfloor n/2 \rfloor} \binom{n}{k} e^{i(n-2(n-k))\theta} \right) \\[1em]
&= \frac{1}{2^n} \sum_{j=0}^{\lfloor n/2 \rfloor} \binom{n}{j} \left(e^{i(n-2j)\theta} + e^{-i(n-2j)\theta} \right)
\end{aligned}
$$

$$= \frac{1}{2^{n-1}} \sum_{j=0}^{\lfloor n/2 \rfloor} \binom{n}{j} \cos((n - 2j)\theta).$$

The derivations for the other cases are similar (Exercise 6).

Example 7.14. Using Equation (7.35) we have

$$\cos^4(x) = 1/8 \, \cos(4\,x) + 1/2 \, \cos(2\,x) + 3/8. \qquad \square$$

Procedures for trigonometric contraction are given in Figures 7.6, 7.7, and 7.8. The recursion that is used to traverse all operands of an expression tree is obtained with the *Map* operator in line 4 of *Contract_trig*. At line 6, this procedure calls on *Contract_trig_rules*, which applies algebraic expansion and calls on other procedures that apply the transformation rules in (7.30), (7.31), and (7.32).

The second type of recursion occurs when either algebraic expansion or an application of one of the contraction rules creates a new sum, product, or power that is not in contracted form. This recursion is invoked in *Contract_trig_rules* at line 15 through *Contract_trig_product* and directly at line 21. At line 1 of *Contract_trig_rules*, we algebraically expand the

---

Procedure *Contract_trig(u)*;
Input
    u : an algebraic expression;
Output
    an algebraic expression in trigonometric-contracted form;
Local Variables
    v;
Begin
1   **if** *Kind(u)* ∈ {**integer, fraction, symbol**} **then**
2   *Return(u)*
3   **else**
4   *v := Map(Contract_trig, u)*;
5   **if** *Kind(v)* ∈ {" ∗ "," ∧ " } **then**
6   *Return(Contract_trig_rules(v))*
7   **else**
8   *Return(v)*
End

Figure 7.6. The main MPL procedure that transforms an algebraic expression to trigonometric-contracted form. (Implementation: Maple (txt), Mathematica (txt), MuPAD (txt).)

Procedure   *Contract_trig_rules(u)*;
Input
    u : an algebraic expression (sum, product, power) that is sent by
        either *Contract_trig*, *Contract_trig_power*, or a recursive call
        of this procedure;
Output
    an algebraic expression in trigonometric-contracted form;
Local Variables
    v, s, c, d, i, y;
Begin

1	$v := Expand\_main\_op(u)$;
2	**if** $Kind(v) = " \wedge "$ **then**
3	$Return(Contract\_trig\_power(v))$
4	**elseif** $Kind(v) = " * "$ **then**
5	$s := Separate\_sin\_cos(v)$;
6	$c := Operand(s, 1)$;
7	$d := Operand(s, 2)$;
8	**if** $d = 1$ **then**
9	$Return(v)$
10	**if** $Kind(d) \in \{\sin, \cos\}$ **then**
11	$Return(v)$
12	**elseif** $Kind(d) = " \wedge "$ **then**
13	$Return(Expand\_main\_op(c * Contract\_trig\_power(d)))$
14	**else**
15	$Return(Expand\_main\_op(c * Contract\_trig\_product(d)))$
16	**elseif** $Kind(v) = " + "$ **then**
17	$s := 0$;
18	**for** $i := 1$ **to** $Number\_of\_operands(v)$ **do**
19	$y := Operand(v, i)$;
20	**if** $Kind(y) \in \{" * ", " \wedge "\}$ **then**
21	$s := s + Contract\_trig\_rules(y)$
22	**else**
23	$s := s + y$;
24	$Return(s)$
25	**else**
26	$Return(v)$

End

Figure 7.7.   The inner MPL procedure for trigonometric contraction. (Implementation: Maple (txt), Mathematica (txt), MuPAD (txt).)

Procedure $Contract\_trig\_product(u)$;

Input
   u: a product of sines, cosines, and positive integer powers
      of sines and cosines;
Output
   the trigonometric-contracted form of u;
Local Variables
   A, B, θ, ϕ;
Begin
1   **if** $Number\_of\_operands(u) = 2$ **then**
2        $A := Operand(u, 1)$;
3        $B := Operand(u, 2)$;
4   **if** $Kind(A) = " \wedge "$ **then**
5           $A := Contract\_trig\_power(A)$;
6           $Return(Contract\_trig\_rules(A * B))$
7   **elseif** $Kind(B) = " \wedge "$ **then**
8           $B := Contract\_trig\_power(B)$;
9           $Return(Contract\_trig\_rules(A * B))$
10   **else**
11          $\theta := Operand(A, 1)$;
12          $\phi := Operand(B, 1)$;
13   **if** $Kind(A) = \sin$ **and** $Kind(B) = \sin$ **then**
14             $Return(\cos(\theta - \phi)/2 - \cos(\theta + \phi)/2)$
15   **elseif** $Kind(A) = \cos$ **and** $Kind(B) = \cos$ **then**
16             $Return(\cos(\theta + \phi)/2 + \cos(\theta - \phi)/2)$
17   **elseif** $Kind(A) = \sin$ **and** $Kind(B) = \cos$ **then**
18             $Return(\sin(\theta + \phi)/2 + \sin(\theta - \phi)/2)$
19   **elseif** $Kind(A) = \cos$ **and** $Kind(B) = \sin$ **then**
20             $Return(\sin(\theta + \phi)/2 + \sin(\phi - \theta)/2)$
21   **else**
22          $A := Operand(u, 1)$;
23          $B := Contract\_trig\_product(u/A)$;
24          $Return(Contract\_trig\_rules(A * B))$
      End

Figure 7.8. The MPL procedure $Contract\_trig\_product$ that contracts products whose operands are sines, cosines, or positive integer powers of sines or cosines. (Implementation: Maple (txt), Mathematica (txt), MuPAD (txt).)

input expression using the operator $Expand\_main\_op$ (Exercise 6, page 258). When v is a power (lines 2-3), we contract using $Contract\_trig\_power$ which checks if v is a positive integer power of a sine or cosine and if so, applies Equation (7.35) or Equation (7.36) (Exercise 7).

Next, in lines 4-15, when v is a product we first apply *Separate_sin_cos* (Exercise 12, page 152) which returns a two element list with the operands of v separated into two categories: the product of the operands that are sines, cosines, or positive integer powers of sines and cosines (represented by d), and the product of the remaining operands (represented by c). At lines 8-9, when $d = 1$, there are no opportunities for contraction, and so v is returned. In a similar way, at lines 10-11, when d is a sine or cosine, there are no opportunities for contraction, and so v is returned. At lines 12-13, when d is a positive integer power of a sine or cosine, we contract using *Contract_trig_power* which applies Equation (7.35) or Equation (7.36) (Exercise 7). Because this procedure returns a sum, we algebraically expand to distribute c over the sum so that property (3) in Definition 7.11 is satisfied. Line 15 handles the case when d is a product of sines, cosines, or positive integer powers of sines and cosines using the procedure *Contract_trig_product* which is described below. Again, expansion is required because the output of this procedure is a sum.

Lines 16-24 handle the case when v is a sum, and lines 25-26 handle other types of expressions that may arise because of the expansion at line 1.

Contract_trig_product, which contracts a product of sines, cosines and positive integer powers of sines and cosines, is shown in Figure 7.8. The case where u has two operands is handled in lines 1-20. When one of the operands is a power, this power is contracted (line 5 or 8), and the new product is contracted with a recursive call to *Contract_trig_rules* (line 6 or 9). At lines 11-20, both A and B are either sines or cosines, and so we apply the transformations in (7.30), (7.31), or (7.32). The case when u has three or more operands is handled in lines 21-24. In this situation, the product with the first operand removed is contracted recursively (line 23), and then the new product is contracted with *Contract_trig_rules* (line 24).

Simplification of Trigonometric Expressions

We now have all the building blocks that are needed to construct an operator that can verify a large class of trigonometric identities.

Automatic Simplification of Trigonometric Functions.
Because our simplification operator performs in the context of automatic simplification, we consider first the trigonometric transformations that are applied in this setting. These include the following transformations.

1. **Evaluation of trigonometric functions.** Let $f(x)$ be a trigonometric function. Typically, automatic simplification evaluates $f(k\,\pi/n)$ where k and $n \neq 0$ are integers and n is small (usually $n = 1, 2, 3, 4, 6$).

Although the values $f(k\pi/n)$ can always be expressed using radicals, the representations are quite involved for large values of n. For example

$$\sin(\pi/60) = \frac{\sqrt{5+\sqrt{5}}}{8} - \frac{\sqrt{5+\sqrt{5}}\sqrt{3}}{8} - \frac{\left(-\frac{\sqrt{5}}{4}+1/4\right)\sqrt{2}}{4}$$

$$- \frac{\left(-\frac{\sqrt{5}}{4}+1/4\right)\sqrt{2}\sqrt{3}}{4}. \tag{7.37}$$

Because it is rarely useful to evaluate $f(k\pi/n)$ for large values of n, these evaluations are usually not performed during automatic simplification (Exercise 3, page 56).

2. **Transformation to argument with positive sign and other standard forms.** In most computer algebra systems, automatic simplification transforms a trigonometric function to an equivalent form with an argument with positive sign. This includes transformations such as

$$\sin(-2/3) \rightarrow -\sin(2/3), \quad \sin(-x) \rightarrow -\sin(x), \quad \cos(-2\,a\,b) \rightarrow \cos(2\,a\,b).$$

In addition, in both Maple and Mathematica automatic simplification transforms trigonometric functions so that arguments that are sums are transformed to a standard form, although each system uses its own scheme to determine the standard form. For example, the Maple system obtains the transformation

$$\sin(1 - \mathbf{x}) \rightarrow -\sin(\mathbf{x} - 1),$$

while the Mathematica system obtains the opposite transformation

$$\mathrm{Sin}[\mathbf{x} - 1] \rightarrow -\mathrm{Sin}[1 - \mathbf{x}].$$

3. **Transformations to arguments in the first quadrant.** In some systems, automatic simplification transforms a trigonometric function with an argument that includes a rational multiple of π to an equivalent function where the multiple of π is between 0 and $\pi/2$. Typical transformations are

$$\sin(15\pi/16) \rightarrow \sin(\pi/16), \quad \sin(x + 2\pi/3) \rightarrow \cos(x + \pi/6).$$

4. **Elementary trigonometric expansions.** In some systems, automatic simplification applies a limited form of trigonometric expansion when the argument of a trigonometric function is a sum with an operand of the form $k\pi/2$ where k is an integer. For example,

$$\sin(x + \pi/2 + y) \rightarrow \cos(x + y), \quad \cos(x + 2\pi) \rightarrow \cos(x).$$

MPL	Maple	Mathematica	MuPAD
$\sin(\pi/3)$ $\to \sqrt{3}/2$	$\sqrt{3}/2$	$\sqrt{3}/2$	$\sqrt{3}/2$
$\sin(-x)$ $\to -\sin(x)$	$-\sin(x)$	$-\text{Sin}[x]$	$-\sin(x)$
$\sin(1-x)$ $\to -\sin(x-1)$	$-\sin(x-1)$	$\text{Sin}[1-x]$	$\sin(-x+1)$
$\sin(-1+x)$ $\to \sin(-1+x)$	$\sin(-1+x)$	$-\text{Sin}[1-x]$	$\sin(x-1)$
$\sin(15\,\pi/16)$ $\to \sin(\pi/16)$	$\sin(\pi/16)$	$\text{Sin}[15\pi/16]$	$\sin(\pi/16)$
$\sin(x+2\pi/3)$ $\to \cos(x+\pi/6)$	$\cos(x+\pi/6)$	$\text{Sin}[2\,\pi/3+x]$	$\sin(x+2\pi/3)$
$\sin(x+\pi/2+y)$ $\to \cos(x+y)$	$\cos(x+y)$	$\text{Cos}[x+y]$	$\sin(x+y+\pi/2)$
$\cos(x+2\pi)$ $\to \cos(x)$	$\cos(x)$	$\text{Cos}[x]$	$\cos(x)$
$\sin(x)/\cos(x)$ $\to \sin(x)/\cos(x)$	$\sin(x)/\cos(x)$	$\text{Tan}[x]$	$\sin(x)/\cos(x)$

Figure 7.9. Examples of trigonometric transformations in automatic simplification in Maple, Mathematica, and MuPAD. (Implementation: Maple (mws), Mathematica (nb), MuPAD (mnb).)

However, for other rational multiples of π (such as $\sin(x + \pi/6)$), the expansion does not occur.

5. **Function transformations.** The Mathematica system obtains the following function transformations in automatic simplification:

$$\text{Sin}[x]/\text{Cos}[x] \;\; \to \;\; \text{Tan}[x],$$
$$\text{Cos}[x]/\text{Sin}[x] \;\; \to \;\; \text{Cot}[x],$$
$$1/\text{Sin}[x] \;\; \to \;\; \text{Csc}[x],$$
$$1/\text{Cos}[x] \;\; \to \;\; \text{Sec}[x].$$

These transformation are not obtained by automatic simplification in either Maple or MuPAD.

Examples of trigonometric transformations in automatic simplification in Maple, Mathematica, and MuPAD are given in Figure 7.9.

The Simplification Algorithm. To motivate the simplification algorithm, let's consider a number of examples.

Example 7.15. Consider the expression

$$(\cos(x) + \sin(x))^4 + (\cos(x) - \sin(x))^4 + \cos(4x) - 3. \qquad (7.38)$$

The trigonometric contraction algorithm simplifies this expression to 0.

On the other hand, the trigonometric expansion algorithm described in Section 7.1 (together with algebraic expansion) does not simplify Expression (7.38) to 0. These operations obtain the trigonometric-expanded form

$$3 \cos^4(x) + 6 \cos^2(x) \sin^2(x) + 3 \sin^4(x) - 3. \qquad \square$$

Example 7.16. Consider the expression

$$\sin(x) + \sin(y) - 2 \sin(x/2 + y/2) \cos(x/2 - y/2).$$

Again, the contraction algorithm simplifies this expression to 0. On the other hand, trigonometric expansion gives the expanded form

$$\sin(x) + \sin(y) - \Big(2 \left(\sin(x/2) \cos(y/2) + \cos(x/2) \sin(y/2) \right)$$

$$(\cos(x/2) \cos(y/2) + \sin(x/2) \sin(y/2)) \Big). \qquad \square$$

These examples suggest that trigonometric contraction is a more powerful simplifier than trigonometric expansion. However, the next example shows that both expansion and contraction play a role in simplification.

Example 7.17. Consider the expression

$$\sin^3(x) + \cos^3(x + \frac{\pi}{6}) - \sin^3(x + \frac{\pi}{3}) + \frac{3 \sin(3x)}{4}. \qquad (7.39)$$

Although this expression simplifies to 0, this is not obtained with trigonometric contraction, which obtains

$$\frac{3}{4} \sin(x) + \frac{3}{4} \cos \left(x + \frac{\pi}{6} \right) - \frac{3}{4} \sin \left(x + \frac{\pi}{3} \right).$$

The problem here is that the simplification requires the trigonometric expansion of $\cos(x + \pi/6)$ and $\sin(x + \pi/3)$ and evaluation of the resulting sin and cos functions at $\pi/3$ or $\pi/6$. The simplification to 0 is obtained by first expanding Expression (7.39) and then contracting the resulting expression.
$\square$

Example 7.18. Consider the identity

$$\frac{\sin(x) + \sin(3x) + \sin(5x) + \sin(7x)}{\cos(x) + \cos(3x) + \cos(5x) + \cos(7x)} = \tan(4x).$$

Procedure *Simplify_trig*(*u*);
Input
    u : an algebraic expression;
Output
    either an algebraic expression in trigonometric contracted form
    or the global symbol **Undefined**;
Local Variables
    v, *w*, *n*, *d*;
Begin
1   *v* := *Trig_substitute*(*u*);
2   *w* := *Rationalize_expression*(*v*);
3   *n* := *Expand_trig*(*Numerator*(*w*));
4   *n* := *Contract_trig*(*n*);
5   *d* := *Expand_trig*(*Denominator*(*w*));
6   *d* := *Contract_trig*(*d*);
7   **if** *d* = 0 **then**
8   *Return*(**Undefined**)
9   **else**
10   *Return*(*n*/*d*)
End

Figure 7.10. The MPL procedure *Simplify_trig*. (Implementation: Maple (txt), Mathematica (txt), MuPAD (txt).)

We verify the identity by subtracting the right side from the left side and showing this expression simplifies to 0. Our algorithm does this by replacing $\tan(4x)$ with $\sin(4x)/\cos(4x)$, rationalizing the resulting expression, and then contracting the numerator of the rationalized form. In this case the numerator of the rationalized expression is

$$\cos(4\,x)\,(\sin(x) + \sin(3\,x) + \sin(5\,x) + \sin(7\,x))$$
$$- (\cos(x) + \cos(3\,x) + \cos(5\,x) + \cos(7\,x))\sin(4\,x),$$

which has 0 as a contracted form. □

A procedure that obtains the simplifications in the above examples is given in Figure 7.10. At line 1 we form a new expression by replacing the tan, sec, cot, and csc functions with equivalent forms with sin and cos using the *Trig_substitute* operator[9] (given in Section 5.2), and at line 2 we ratio-

---

[9] These substitutions do not occur in Mathematica because automatic simplification performs the inverse transformations (see footnote 2, page 187). Our implementation of the algorithm in this system does not include the step in line 1.

nalize the resulting expression. At line 3, we trigonometrically expand the numerator of w so that a rational multiple of π that appears as an operand of a sum in an argument of a sin or cos now appears directly as the argument of a sin or cos. This operation together with automatic simplification obtains the numerical representation of some sines and cosines. (This operation is required in Example 7.17.) At line 4, we contract the numerator and in lines 5 and 6 apply expansion and contraction to the denominator. Finally, at line 7 we check if the denominator has been simplified to 0, and, if so, return the global symbol **Undefined**. Otherwise, we return n/d.

Appraisal of *Simplify_trig*. The *Simplify_trig* operator can verify many trigonometric identities that appear in trigonometry textbooks. In this role, it is most effective by showing that the difference of the two sides of an identity simplify to 0, as was done in Example 7.18. For expressions that do not simplify to 0, however, it is less successful. For example, for the identity

$$\frac{\sin(x) + \sin(3\,x) + \sin(5\,x) + \sin(7\,x)}{\cos(x) + \cos(3\,x) + \cos(5\,x) + \cos(7\,x)} = \frac{\sin(4\,x)}{\cos(4\,x)},$$

the smaller expression on the right is a simpler form than the one on the left. However, *Simplify_trig* does not change either side of this expression because the numerators and denominators on both sides are in contracted form.

In addition, identities that require other (non-trigonometric) transformations might not be verified with *Simplify_trig*. For example, although the expression

$$\sin^2\left(\frac{x+1}{x+2}\right) + \cos^2\left(\frac{1+1/x}{1+2/x}\right)$$

simplifies to 1, this is not obtained by our algorithm because it does not recognize the arguments of sin and cos as equivalent expressions.

The algorithm depends, of course, on the transformations applied to trigonometric functions during automatic simplification. For example, both the Maple and Mathematica implementations of the algorithm simplify Expression (7.39) to 0. On the other hand, the MuPAD implementation obtains

$$-\frac{\sin(x-y)}{2} - \frac{\sin(-x+y)}{2}$$

because arguments of sines that are sums are not transformed to a standard form in this system. In a similar way, both the Maple and MuPAD implementations simplify

$$\frac{\sin(x) + \sin(3\,x) + \sin(5\,x) + \sin(7\,x)}{\cos(x) + \cos(3\,x) + \cos(5\,x) + \cos(7\,x)} - \tan(4\,x)$$

to 0, while the Mathematica implementation does not because automatic simplification does not permit the transformation

$$\tan(4\,x) \rightarrow \sin(4\,x)/\cos(4\,x).$$

Exercises

1. Determine if each of the following expressions is in exponential-contracted form. If the expression is not in exponential-contracted form, transform it to this form.

   (a) $\dfrac{1}{(1+\exp(x))(1+\exp(y))}$.

   (b) $\exp((x+y)(x-y))$.

   (c) $\exp((a+b)\exp(x))^{\exp(y)}$.

2. An algebraic expression u is in *log-contracted* form if it satisfies the following two properties.

   (a) A sum in u has at most one operand that is a logarithm.

   (b) A product in u that has an operand that is a logarithm does not also have an operand that is an integer or fraction.

   For example, $a\ln(x) + a\ln(y) + (\ln(x))^2$ is in log-contracted form, while $\ln(x) + \ln(y) + 2\ln(x)$ is not. An algebraic expression can be transformed to log-contracted form by applying the transformations

   $$\ln(u) + \ln(v) \;\rightarrow\; \ln(u\,v), \tag{7.40}$$
   $$n\ln(u) \;\rightarrow\; \ln(u^n), \tag{7.41}$$

   where n is an integer or fraction. If a product contains more than one logarithm, (7.41) is applied to the first operand that is a logarithm. Give a procedure $Contract\_log(u)$ that transforms an algebraic expression u to log-contracted form.

3. The $Contract\_exp$ algorithm encounters a division by 0 if the transformation rules transform a sub-expression in a denominator to 0. For example, this occurs with $1/(\exp(2\,x) - (\exp(x))^2)$. Modify the algorithm so that it recognizes this situation and returns the global symbol **Undefined** when it occurs.

4. Let u be an algebraic expression. Give a procedure for $Simplify\_exp(u)$ which rationalizes u and then exponentially contracts the numerator and denominator of the resulting expression. If the denominator contracts to 0, the procedure returns the global symbol **Undefined**. Your procedure should simplify Expression (7.29) to 0.

5. Find the trigonometric-contracted form of $\sin(x)\cos^2(x)\cos(2\,x)$.

6. Verify Equation (7.35) for n even and Equation (7.36) for n even and n odd.

7. Let u be a power. Give a procedure $Contract\_trig\_power(u)$ that does the following.

   (a) If the exponent of u is a positive integer and the base is a sine or cosine, then contract u using Equation (7.35) or Equation (7.36). The floor function is obtained in Maple and MuPAD with `floor` and in Mathematica with `Floor`.

   (b) If the exponent of u is not a positive integer or the base is not a sine or cosine, then return u.

8. The $Contract\_trig$ algorithm encounters a division by 0 if the transformation rules transform a sub-expression in a denominator to 0. For example, this occurs with $1/(\sin(2\,x) - 2\sin(x)\cos(x))$. Modify the algorithm so that it recognizes this situation and returns the global symbol **Undefined** when it occurs.

9. (a) State and derive the formulas similar to Equation (7.35) or Equation (7.36) for $\sinh^n(\theta)$ and $\cosh^n(\theta)$.

   (b) Modify the algorithm $Contract\_trig$ so that it also contracts the hyperbolic functions sinh and cosh.

   Identities for sinh and cosh are given in Exercise 9, page 287.

10. This exercise refers to the $Trig\_form$ operator described in Exercise 8, page 211. Modify this procedure so that it evaluates $\int \sin^m(a\,x)\cos^n(a\,x)\,dx$ by transforming $\sin^m(a\,x)\cos^n(a\,x)$ to contracted form and then applying the $Integral$ operator to the contracted form.

11. Each of the following expressions simplifies to 0. Is this simplification obtained by $Simplify\_trig$? If 0 is obtained, then explain how this is done. If not, explain why not.

   (a) $\tan(x/2) - (\sin(x)/(1 + \cos(x)))$.

   (b) $\sin^2(15\pi/16) + \cos^2(\pi/16) - 1$.

   (c) $\sin^2(x^2 - 1) + \cos^2(1 - x^2) - 1$.

   (d) $\sin^2\left(x + \dfrac{1}{x}\right) + \cos^2\left(\dfrac{x^2 + 1}{x}\right) - 1$.

   (e) $\sin\left(x\left(1 + \dfrac{\pi}{6\,x}\right)\right) - \sqrt{3}/2\,\sin(x) - 1/2\,\sin(x)$.

   (f) $\sin\left((\sin^2(x) + \cos^2(x))\,(x + \pi/6)\right) - \sqrt{3}/2\,\sin(x) - 1/2\,\sin(x)$.

   (g) $8\cos^3(2\pi/7) + 4\cos^2(2\pi/7) - 4\cos(2\pi/7) - 1$.

Further Reading

See Hobson [47] for other approaches to trigonometric expansion and contraction. Gutierrez and Recio [42] discuss new algorithms for trigonometric simplification and some applications to robotics.

Bibliography

[1] Williams W. Adams and Philippe Loustaunau. *An Introduction to Gröbner Bases*. Graduate Studies in Mathematics, Volume 3. American Mathematical Society, Providence, RI, 1994.

[2] Alkiviadis G. Akritas. *Elements of Computer Algebra with Applications*. John Wiley & Sons, New York, 1989.

[3] Michael Artin. *Algebra*. Prentice Hall, Inc., Englewood Cliffs, NJ, 1991.

[4] E. J. Barbeau. *Polynomials*. Springer-Verlag, New York, 1989.

[5] David Barton and Richard Zippel. Polynomial decomposition algorithms. *Journal of Symbolic Computation*, 1(1):159–168, 1985.

[6] Thomas Becker, Volker Weispfenning, and Heinz Kredel. *Gröbner bases, A Computational Approach to Commutative Algebra*. Springer-Verlag, New York, 1993.

[7] Laurent Bernardin. A review of symbolic solvers. *SIGSAM Bulletin*, 30(1):9–20, March 1996.

[8] Laurent Bernardin. A review of symbolic solvers. In Michael J. Wester, editor, *Computer Algebra Systems, A Practical Guide*, pages 101–120. John Wiley & Sons, Ltd., New York, 1999.

[9] A. S. Besicovitch. On the linear independence of fractional powers of integers. *J. London Math. Soc.*, 15:3–6, 1940.

[10] Garrett Birkhoff and Saunders Mac Lane. *A Survey of Modern Algebra*. A K Peters, Ltd., Natick, MA, 1997.

[11] E. Bond, M. Auslander, S. Grisoff, R. Kenney, M. Myszewski, J. Sammet, R. Tobey, and S. Zilles. Formac–An experimental formula manipulation compiler. In *Proc. 19th ACM National Conference*, pages K2.1–1–K2.1–11, August 1964.

[12] William E. Boyce and Richard C. DiPrima. *Elementary Differential Equations*. Sixth Edition. John Wiley & Sons, New York, 1997.

[13] Manuel Bronstein. *Symbolic Integration I, Transcendental Functions*. Springer-Verlag, New York, 1997.

[14] W. S. Brown. On Euclid's algorithm and the computation of polynomial greatest common divisors. *Journal of the Association for Computing Machinery*, 18(4):478–504, October 1971.

[15] W. S. Brown. The subresultant prs algorithm. *ACM Transactions on Math. Software*, 4(3):237–249, September 1978.

[16] W. S. Brown and J. F. Traub. On Euclid's algorithm and the theory of subresultants. *Journal of the Association for Computing Machinery*, 18(4):505–514, October 1971.

[17] B. Buchberger, G. E. Collins, R. Loos, and R. Albrecht. *Computer Algebra, Symbolic and Algebraic Computation*. Second Edition. Springer-Verlag, New York, 1983.

[18] Bruno Buchberger. Gröbner bases: An algorithmic method in polynomial ideal theory. In N. K. Bose, editor, *Recent Trends in Multidimensional Systems Theory*, pages 184–232. D. Reidel Publishing Company, Dordrecht, Holland, 1985.

[19] Ronald Calinger. *Classics of Mathematics*. Moore Publishing Company Inc., Oak Park, IL, 1982.

[20] Shang-Ching Chou. *Mechanical Geometry Theorem Proving*. D. Reidel Publishing Company, Boston, 1988.

[21] Barry A. Cipra. Do mathematicians still do math. *Science*, 244:769–770, May 19, 1989.

[22] Henri Cohen. *A Course in Computational Algebraic Number Theory*. Springer-Verlag, New York, 1993.

[23] J. S. Cohen, L. Haskins, and J. P. Marchand. Geometry of equilibrium configurations in the ising model. *Journal of Statistical Physics*, 31(3):671–678, June 1983.

[24] Joel S. Cohen. *Computer Algebra and Symbolic Computation: Mathematical Methods*. A K Peters, Natick, MA, 2002.

[25] George Collins. Subresultants and reduced polynomial remainder sequences. *J. ACM*, 14:128–142, January 1967.

[26] George Collins. The calculation of multivariate polynomial resultants. *J.ACM*, 18(4):515–532, October 1971.

[27] George Collins. Computer algebra of polynomials and rational functions. *American Mathematical Monthly*, 80(7):725–754, September 1973.

[28] Thomas H. Cormen, Charles E. Leiserson, and Ronald L. Rivest. *Introduction to Algorithms*. McGraw-Hill, New York, 1989.

[29] J. H. Davenport, Y. Siret, and E. Tournier. *Computer Algebra, Systems and Algorithms for Algebraic Computation*. Academic Press, New York, 1988.

[30] P. J. Davis and R. Hersh. *The Mathematical Experience*. Birkhäuser, Boston, MA, 1981.

[31] Richard A. Dean. *Classical Abstract Algebra*. Harper and Row, New York, 1990.

[32] William R. Derrick and Stanley I. Grossman. *Differential Equations with Applications*. Third Edition. West Publishing Company, St. Paul, MN, 1987.

[33] F. Dorey and G. Whaples. Prime and composite polynomials. *Journal of Algebra*, 28:88–101, 1974.

[34] H. T. Engstrom. Polynomial substitutions. *Amer. J. of Mathematics*, 63:249–255, 1941.

[35] James F. Epperson. *An Introduction to Numerical Methods and Analysis*. John Wiley & Sons, New York, 2002.

[36] R. J. Fateman. Macsyma's general simplifier. philosophy and operation. In V.E. Lewis, editor, *Proceedings of MACSYMA's Users' Conference*, Washington, D.C., June 20–22 1979, pages 563–582. MIT Laboratory for Computer Science, Cambridge, MA, 1979.

[37] Richard J. Fateman. Symbolic mathematics system evaluators. In Michael J. Wester, editor, *Computer Algebra Systems, A Practical Guide*, pages 255–284. John Wiley & Sons, Ltd., New York, 1999.

[38] Richard J. Gaylord, N. Kamin, Samuel, and Paul R. Wellin. *An Introduction to Programming with Mathematica, Second Edition.* Springer-Verlag, New York, 1996.

[39] K.O. Geddes, S.R. Czapor, and G. Labahn. *Algorithms for Computer Algebra.* Kluwer Academic Publishers, Boston, 1992.

[40] Jürgen Gerhard, Walter Oevel, Frank Postel, and Stefan Wehmeier. *MuPAD Tutorial, English Edition.* Springer-Verlag, New York, 2000.

[41] John W. Gray. *Mastering Mathematica, Programming Methods and Applications.* Second Edition. Academic Press, New York, 1997.

[42] J. Gutierrez and T. Recio. Advances on the simplification of sine-cosine equations. *Journal of Symbolic Computation*, 26(1):31–70, July 1998.

[43] G. H. Hardy and E. M. Wright. *An Introduction To The Theory of Numbers.* Oxford at The Clarendon Press, London, 1960.

[44] K. M. Heal, M. L. Hansen, and K. M. Rickard. *Maple 6 Learning Guide.* Waterloo Maple Inc., Waterloo, ON, Canada, 2000.

[45] André Heck. *Introduction to Maple.* Second Edition. Springer-Verlag, New York, 1996.

[46] I. N. Herstein. *Topics in Algebra.* Second Edition. Xerox Publishing Company, Lexington, MA, 1975.

[47] E. W. Hobson. *Treatise on Plane and Advanced Trigonometry.* Seventh Edition. Dover Publications, Inc., New York, 1957.

[48] Douglas R. Hofstadter. *Gödel, Escher, Bach: An Eternal Golden Braid.* Random House Inc., New York, 1980.

[49] David J. Jeffrey and Albert D. Rich. Simplifying square roots of square roots by denesting. In Michael J. Wester, editor, *Computer Algebra Systems, A Practical Guide*, pages 61–72. John Wiley & Sons, Ltd., New York, 1999.

[50] Richard D. Jenks and Robert S. Sutor. *Axiom, The Scientific Computation System.* Springer-Verlag, New York, 1992.

[51] N. Kajler, editor. *Computer-Human Interaction in Symbolic Computation*. Springer-Verlag, New York, 1998.

[52] Israel Kleiner. Field theory, from equations to axiomatization. part i. *American Mathematical Monthly*, 106(7):677–684, August-September 1999.

[53] Israel Kleiner. Field theory, from equations to axiomatization. part ii. *American Mathematical Monthly*, 106(9):859–863, November 1999.

[54] Morris Kline. *Mathematics and The Search for Knowledge*. Oxford University Press, New York, 1985.

[55] D. Knuth. *The Art of Computer Programming*, volume 2. Second Edition. Addison-Wesley, Reading, MA, 1981.

[56] Donald Knuth, Ronald Graham, and Oren Patashnik. *Concrete Mathematics, A Foundation For Computer Science*. Addison-Wesley, Reading, MA, 1989.

[57] K. Korsvold. An on-line algebraic simplify program. Technical report, Stanford University, 1965. Stanford University Artificial Intelligence Project, Memorandum 37.

[58] J. S. Kowalik, editor. *Coupling Symbolic and Numerical Computing in Expert Systems*. Elsevier Science Publishers, New York, 1986.

[59] Dexter Kozen and Susan Landau. Polynomial decomposition algorithms. *Journal of Symbolic Computation*, 7:445–456, 1989.

[60] Susan Landau. Simplification of nested radicals. *SIAM J. Comput.*, 21(1):85–110, February 1992.

[61] Ulrich Libbrecht. *Chinese Mathematics in the Thirteenth Century*. MIT Press, Cambridge, MA, 1973.

[62] R. Lidl and H. Niederreiter. *Introduction to Finite Fields and their Applications*. Revised Edition. Cambridge University Press, New York, 1994.

[63] C. C. Lin and L. A. Segel. *Mathematics Applied to Deterministic Problems in the Natural Sciences*. Classics in Applied Mathematics 1. Society for Industrial and Applied Mathematics, Philadelphia, 1988.

[64] John D. Lipson. *Elements of Algebra and Algebraic Computing*. Benjamin/Cummings, Menlo Park, CA, 1981.

[65] Stephen B. Maurer and Anthony Ralston. *Discrete Algorithmic Mathematics*. A K Peters, Ltd., Natick, MA, 1998.

[66] Maurice Mignotte. *Mathematics for Computer Algebra*. Springer-Verlag, New York, 1991.

[67] Maurice Mignotte and Doru Ştefănescu. *Polynomials, An Algorithmic Approach*. Springer-Verlag, New York, 1999.

[68] Bhubaneswar Mishra. *Algorithmic Algebra*. Springer-Verlag, New York, 1993.

[69] M. B. Monagan, K. O. Geddes, K. M. Heal, G. Labahn, S. M. Vorkoetter, and J. McCarron. *Maple 6 Programming Guide*. Waterloo Maple Inc., Waterloo, ON, Canada, 2000.

[70] Joel Moses. *Symbolic Integration*. PhD thesis, MIT, September 1967.

[71] Joel Moses. Algebraic simplification: A guide for the perplexed. *Communications of the ACM*, 14(8):527–537, August 1971.

[72] George M. Murphy. *Ordinary Differential Equations and Their Solutions*. D. Van Nostrand, New York, 1960.

[73] David Musser. *Algorithms for Polynomial Factorization*. PhD thesis, Department of Computer Science, University of Wisconsin, 1971.

[74] Paul J. Nahin. *An Imaginary Tale, The Story of* $\sqrt{-1}$. Princeton University Press, Princeton, NJ, 1998.

[75] Jurg Nievergelt, J. Craig Farrar, and Edward M. Reingold. *Computer Approaches to Mathematical Problems*. Prentice-Hall, Englewood Cliffs, NJ, 1974.

[76] F. S. Nowlan. Objectives in the teaching college mathematics. *American Mathematical Monthly*, 57(1):73–82, February 1950.

[77] R. Pavelle, M. Rothstein, and J. P. Fitch. Computer algebra. *Scientific American*, 245:136–152, 1981.

[78] Louis L. Pennisi. *Elements of Complex Variables*. Holt, Rinehart and Winston, New York, 1963.

[79] Charles Pinter. *A Book of Abstract Algebra*. Second Edition. McGraw-Hill, New York, 1990.

[80] Marcelo Polezzi. A geometrical method for finding an explicit formula for the greatest common divisor. *American Mathematical Monthly*, 104(5):445–446, May 1997.

[81] Frank Postel and Paul Zimmermann. Solving ordinary differential equations. In Michael J. Wester, editor, *Computer Algebra Systems, A Practical Guide*, pages 191–209. John Wiley & Sons, Ltd., New York, 1999.

[82] T. W. Pratt. *Programming Languages, Design and Implementation.* Second Edition. Prentice Hall, Englewood Cliffs, NJ, 1984.

[83] Gerhard Rayna. *Reduce, Software for Algebraic Computation.* Springer-Verlag, New York, 1987.

[84] D. Richardson. Some undecidable problems involving elementary functions of a real variable. *Journal of Symbolic Logic*, 33(4):511–520, December 1968.

[85] J. F. Ritt. Prime and composite polynomials. *Trans. Am. Math. Soc.*, 23:51–66, 1922.

[86] P. Sconzo, A. LeSchack, and R. Tobey. Symbolic computation of f and g series by computer. *The Astronomical Journal*, 70(1329):269–271, May 1965.

[87] George F. Simmons. *Differential Equations with Applications and Historical Notes.* Second Edition. McGraw-Hill, New York, 1991.

[88] George F. Simmons. *Calculus with Analytic Geometry.* Second Edition. McGraw-Hill, New York, 1996.

[89] Barry Simon. Symbolic magic. In Michael J. Wester, editor, *Computer Algebra Systems, A Practical Guide*, pages 21–24. John Wiley & Sons, Ltd., New York, 1999.

[90] Barry Simon. Symbolic math powerhouses revisited. In Michael J. Wester, editor, *Computer Algebra Systems, A Practical Guide*. John Wiley & Sons, Ltd., New York, 1999.

[91] Trevor J. Smedley. Fast methods for computation with algebraic numbers. Research Report CS-90-12, Department of Computer Science, University of Waterloo, May 1990.

[92] Jerome Spanier and Keith B. Oldham. *An Atlas of Functions.* Hemisphere Publishing Corporation, New York, 1987.

[93] Frederick W. Stevenson. *Exploring the Real Numbers*. Prentice Hall, Upper Saddle River, NJ, 2000.

[94] David R. Stoutemyer. Crimes and misdemeanors in the computer algebra trade. *Notices of the American Mathematical Society*, 38(7):778–785, September 1991.

[95] R. Tobey, R. Bobrow, and S. Zilles. Automatic simplification in Formac. In *Proc. AFIPS 1965 Fall Joint Computer Conference*, volume 27, pages 37–52. Spartan Books, Washington, DC, November 1965. Part 1.

[96] Joachim von zur Gathen and Jürgen Gerhard. *Modern Computer Algebra*. Cambridge University Press, New York, 1999.

[97] Paul Wang and Barry Trager. New algorithms for polynomial square free decomposition over the integers. *SIAM Journal of Comp.*, 8:300–305, 1979.

[98] Mark Allen Weiss. *Data Structures and Problem Solving Using C++*. Second Edition. Addison-Wesley, Reading, MA, 2000.

[99] Clark Weissman. *Lisp 1.5 Primer*. Dickenson Publishing Company, Belmont, CA, 1967.

[100] Michael J. Wester. *Computer Algebra Systems, A Practical Guide*. John Wiley & Sons, Ltd., New York, 1999.

[101] F. Winkler. *Polynomial Algorithms in Computer Algebra*. Springer-Verlag, New York, 1996.

[102] Stephen Wolfram. *The Mathematica Book*. Fourth Edition. Cambridge University Press., New York, 1999.

[103] D. Wooldridge. An algebraic simplify program in Lisp. Technical report, Stanford University, December 1965. Artificial Intelligence Project, Memo 11.

[104] W. A. Wulf, M. Shaw, P. Hilfinger, and L. Flon. *Fundamental Structures of Computer Science*. Addison-Wesley, Reading, MA, 1981.

[105] Chee Keng Yap. *Fundamental Problems of Algorithmic Algebra*. Oxford University Press, New York, 2000.

[106] David Y. Y. Yun. On square-free decomposition algorithms. In R. D. Jenks, *Proceedings of the 1976 ACM Symposium of Symbolic and Algebraic Computation*, pages 26–35. ACM, New York, 1976.

[107] David Y. Y. Yun and David R. Stoutemyer. Symbolic mathematical computation. In J. Belzer, A.G. Holzman, and A. Kent, editors, *Encyclopedia of Computer Science and Technology*, volume 15, pages 235–310. M. Dekker, New York, 1980.

[108] Richard Zippel. *Effective Polynomial Computation*. Kluwer Academic Publishers, Boston, 1993.

[109] Daniel Zwillinger. *Handbook of Differential Equations*. Academic Press, Boston, MA, 1989.

Index

317